$23.75

OPERATING *the* TRACTOR-LOADER-BACKHOE

By

Gary J. Ober

CRAFTSMAN

Craftsman Book Company

6058 Corte del Cedro, P. O. Box 6500, Carlsbad, California 92008

Photographs used by permission of Ford Motor Company. Use of the photographs is not an endorsement, recommendation or approval of the methods illustrated and/or described herein.

Library of Congress Cataloging-in-Publication Data

Ober, Gary J.
Operating the tractor-loader-backhoe,

Rev. ed. of: Operating techniques for the tractor-loader-backhoe. 1st ed. 1983, c1982.
Includes index.
1. Excavating machinery. I. Ober, Gary J. Operating techniques for the tractor-loader-backhoe. II. Title.
TA735.025 1985 621.8'65 85-29123
ISBN 0-934041-07-5

CONTENTS

PART I

PART II

PART I.

The tractor-loader-backhoe (or "TLB" as I'll call it here) is a highly versatile, productive piece of equipment. It's ideal for most of the trenching that's done on a building site. It's perfect for loading soil and heavy objects into a truck. It handles light demolition well. And, of course, it backfills trenches and does minor grading much faster than could be done by hand. Many builders consider a TLB as essential as a pickup truck or a skilsaw.

Whether you're buying or leasing a TLB or just using a rented machine occasionally, this book has the information you need to get the best results possible from the equipment you're using.

What You Need to Know

This book isn't a technical manual. And it doesn't cover any particular type of TLB. The various manufacturers have all the technical manuals you'll need on the equipment they manufacture and sell. Nor does this book recommend one brand of TLB over any other. They're all good.

What this book does is explain in the most concise, most direct way possible, exactly what you should know to get the most productivity out of the TLB you use. That's not as easy as it sounds. The TLB has considerably more capacity built in than most operators can get out. It will handle types of work that many operators have never tried. And it has safety hazards that can cause serious damage to people or property if the operator's not aware of the potential for accidents.

I've written this book for both the beginning and the experienced operator. It begins with the basics, of course. But there's plenty of useful information in later chapters for even veteran operators. And I've never met an operator who couldn't learn a little more or get rid of a few bad habits. Certainly I haven't stopped learning about the TLB. And I've been an operator for many years.

Steps Toward Your Goal

As you read through the first chapters, you'll notice that my descriptions seem to cover a series of unrelated procedures. Be patient — and learn each step in the order presented. As you read further you'll begin to notice that each procedure becomes a step in a larger process. Skipping one step will make the next more difficult, or impossible.

Later sections of this book emphasize the more difficult job situations: tight work space, working around utilities, precision excavation and the more complicated tasks. Even if you never have a job that demands precise control, you'll learn something by mastering the moves necessary to do precision work. Learn the more demanding tasks and everything else will seem like a piece of cake. That's why I've emphasized techniques and work procedures that cover "worst-case" situations.

Of course, no single book can describe every possible task that a TLB operator is likely to face. Instead, I've distilled what I've learned from several thousand days at the controls of a TLB into a concise, readable guide. Read it and absorb as much as you can — and then refer back to it when you need more specifics for an unusual job.

Get Started Right

A word to beginners: There's no substitute for practice, practice and more practice. Just reading this book isn't enough. It's practicing what you've learned here that makes you an expert. With time, operating a backhoe correctly will become as natural for you as driving a car or tying your shoe.

And don't be discouraged if you have trouble at first. The TLB is among the most difficult pieces of heavy equipment to operate correctly. There are two reasons: First, it's really two machines in one, with separate controls and operating principles for each. Second, the TLB is used for so many tasks that there's a lot to learn. No matter how experienced you are, there's another job yet to be mastered.

As I said, no book will ever be written that covers in detail every possible task a TLB has to perform. But this book will do the next best thing. I'm going to suggest ways for you to find an intelligent, safe, workable solution to any new challenge. Armed with knowledge like that, you're well on the way to becoming a real TLB expert.

The first four chapters of the book cover the basics: machine dimensions, control, safety, and maintenance. This information will be of more interest to beginning operators than to old pros. But don't skip these chapters entirely, even if you've had years on the TLB. There's good advice here that can increase any operator's efficiency.

The chapters on maintenance and safety include some very important information. Don't sit down behind the controls of the TLB if you don't understand what's in these chapters. Any operator who doesn't make safety a top priority and who doesn't have respect for the machine is a hazard to himself and everyone on the job. Reading Chapters 3 and 4 will reduce the chance that you'll have a major, debilitating accident.

These first four chapters are also intended to give beginners a "feel" for the way a TLB responds to control movements. If you've never operated a TLB before, spend some time with Part I before getting in the operator's seat and before going on to Part II.

1. TLB DIMENSIONS

General

The modern tractor-loader-backhoe (usually abbreviated to TLB, but more commonly referred to simply as a backhoe) is an efficient earthmoving machine which combines the features of several types of heavy equipment. It can dig trenches, carry heavy materials in bulk, transport large objects, and work the earth in whatever way required for practically all modern construction, demolition or excavation projects. It has become a common sight on city streets as well as in rural locations.

Like any piece of complicated machinery, the backhoe requires a skilled operator who is proficient in the use, combination and control of the machine's various functions. Becoming a skilled backhoe operator requires thorough understanding of the machine's capabilities and the principles behind its operation, and thousands of hours of practice.

The primary step toward this understanding between man and machine is in learning the TLB's dimensions. The TLB is generally more difficult to maneuver than other types of heavy equipment because it is over-long, over-high, and over-wide. Depending on the make and model, the TLB usually has an overall length of twenty-three feet with the boom drawn in; with the boom extended, the machine may stretch out to a length of thirty-five feet or more. It is at least twelve feet high with the hoe in transport position. And with the two stabilizer arms lowered, the backhoe takes up ten feet from side to side. In short, the TLB can be an awkward machine. Depending on the operator's skill, however, this awkwardness can be minimized and even made to work to his advantage. One way he can do this is by understanding the backhoe's weight distribution.

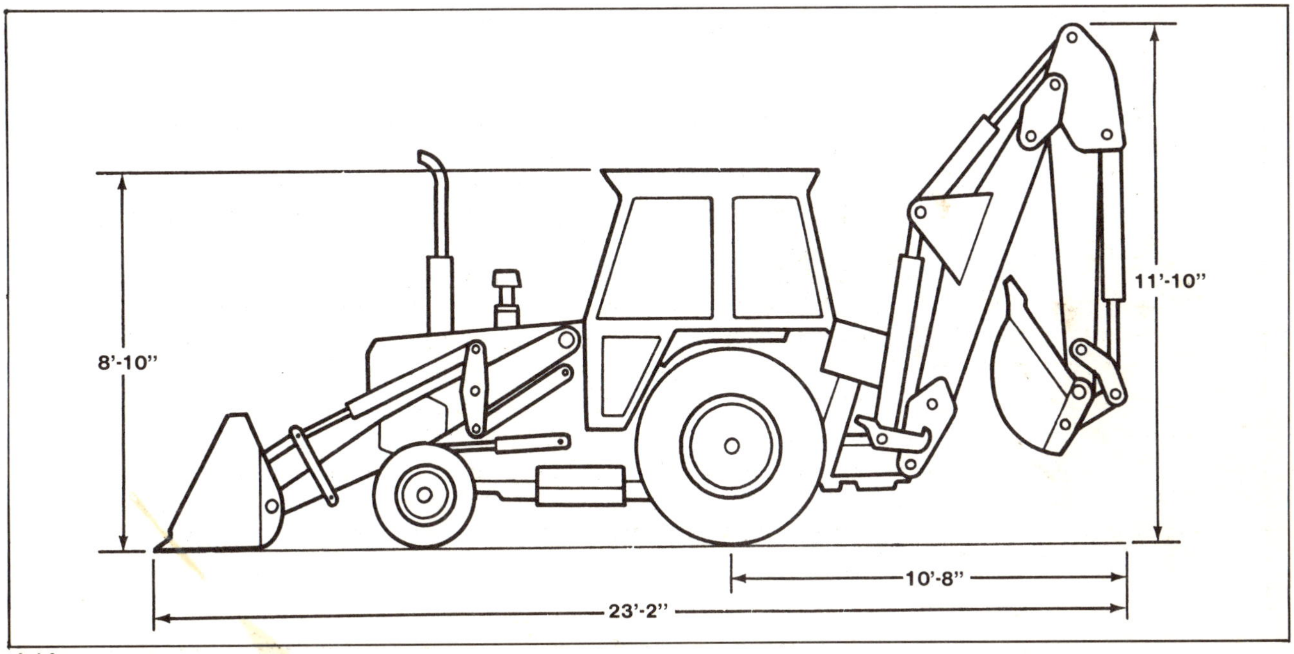

1-1A

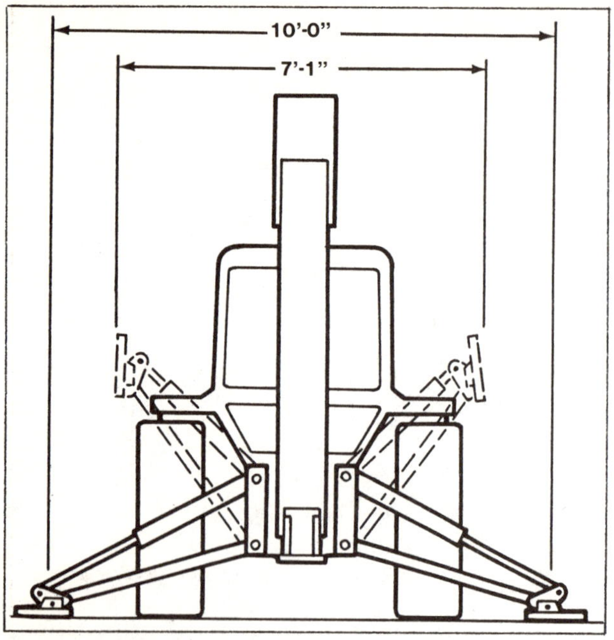

Fig. 1-1 A on previous page shows TLB diagrammed from a side view; Fig. 1-1 B (above) shows the TLB from the rear looking forward. Note width of machine when stabilizers are lowered. Fig. 1-2 (below) shows gross weight of various components of the TLB. Fig. 1-3 (opposite) compares the TLB's center of gravity with that of a skiploader and bulldozer.

Weight Distribution

The TLB consists of three major components: 1) the tractor itself, which houses the diesel-powered engine, the hydraulic pump and the cab; 2) the front loader assembly, and 3) the hoe assembly, mounted on the rear. Because of the flexibility of these components, particulary the hoe itself, how the machine's seven tons are distributed frequently plays a major role in the planning and completion of many jobs.

Approximately one-third of the total weight is in the two-ton hoe assembly, whose main components include the boom, the crowd (or stick), the bucket and the stabilizer units (see diagram). Because of its location high above the ground and far behind the wheelbase, the backhoe's weight transfers the entire machine's center of gravity upward as well as to the rear, as compared to other machines. This, in turn, makes the machine lighter on the front end and somewhat top-heavy. The positioning of this weight when maneuvering around a job can be a constant source of trouble for even an experienced operator, since each job presents different demands and hazards.

GROSS WEIGHT OF TLB COMPONENTS

Loader	Tractor	Backhoe
2,100 lbs.	6,625 lbs.	4,000 lbs.

1-2

CENTER OF GRAVITY COMPARISON

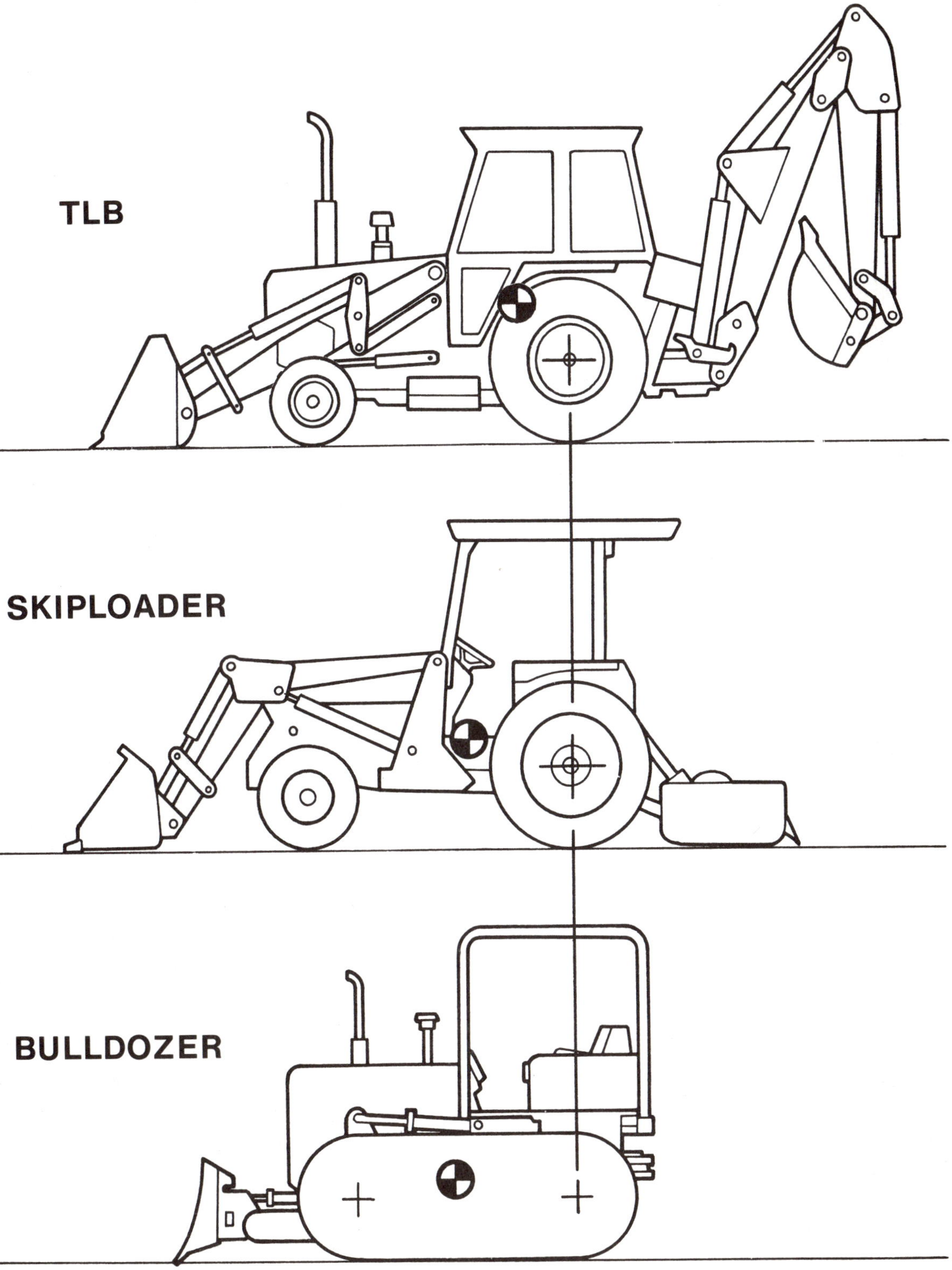

1-3

1-4

1-5

1-6

1-7

1-8

1-9

It is important to note that the TLB's weight distribution does not always stay the same. Moving the machine uphill, for example, shifts the weight to the rear (Fig. 1-4), while going downhill moves it toward the front (Fig. 1-5). When going across a slope, the weight moves to the downhill side of the machine (Fig. 1-6). Raising the loader bucket or extending the hoe moves it to the rear of the machine. Some models have counterweights on the front to prevent this shift and keep the front tires on the ground. While these weights may be effective on flat ground, they are not enough to stabilize the backhoe when operating on a steep grade. (See Fig. 1-7.)

It is vital that the operator understand how these shifts in the machine's weight distribution occur, because whenever it moves beyond the wheelbase, the machine will tip over unless the shift is counteracted by the operator. Fortunately, the backhoe is unique in that the center of gravity can be easily manipulated to the operator's advantage once he learns how to alter the machine's dimensions. Altering the dimensions of the backhoe can turn a heavy, unwieldy piece of equipment into an efficient, relatively compact machine capable of performing a wide variety of tasks.

1-10

1-12

1-11

1-13

To see how the operator solved the awkward problem of transporting this log, turn the page.

Altering Dimensions

In addition to increasing the TLB's efficiency and maneuverability, learning to alter its dimensions is important from a safety standpoint. When crossing a hill, for instance, the machine can become unstable and dangerous to the operator, other workers, property in the area, and even to itself — unless the operator knows how to shift the weight distribution, in this case by swinging the hoe to the uphill side (Fig. 1-8). For steeper hills, the hoe may be extended outward (Fig. 1-9). In cases where the operator is unsure, he can always lower the stabilizer on the downhill side, leaving it just off the ground, until he is sure the machine is stable (Fig. 1-10).

Whenever attempting to climb or descend steep hills, the rear of the machine (the hoe) should face uphill. On steep grades, the hoe should be extended outward, keeping the weight as far back as possible (Fig. 1-11). Otherwise, this weight would tend to be thrown over the machine toward the front. This could lift the rear wheels off the ground. It is important to keep as much weight on the rear wheels as possible, because the brakes are on the rear axle. Reducing the weight on the rear wheels will not only cost the operator traction but also control, two vital elements of operating on grades. On the other hand, going up a hill forward will cause the front tires to leave the ground, again resulting in loss of control. If you must drive up or back down a grade with the hoe on the downward end, stabilize the front wheels by filling the loader bucket with dirt or heavy material (Fig. 1-12).

Knowing how to alter dimensions is important on flat ground as well. An operator will constantly find himself in tight situations with limited working space, requiring maximum maneuverability. In Fig. 1-13, for example, we see an operator transporting a log down a narrow road with the hoe. By swinging the load alongside the machine, however, transport becomes easier (Fig. 1-14). In this case, the road is slightly off-camber, so the operator has stabilized his machine by swinging the log to the uphill side. Also notice that before attempting to transport materials with the hoe, the machine is stabilized by filling the loader bucket (Fig. 1-15).

Another way of shortening the machine to add to its maneuverability is to swing the hoe to the side as shown in Fig. 1-16.

Raising the loader bucket will "shorten" the machine to improve maneuverability, but it also shifts the weight up and to the rear. This should never be attempted on slopes or grades. As a rule of thumb, **keep the loader bucket as low as possible, especially when carrying a load.**

In summary, a skilled operator must know his machine's dimensions. A combination of studying the manufacturer's specifications, and plenty of hands-on practice will show the operator what the machine can and cannot do. Learning how the machine's weight is distributed, and using that knowledge to alter its dimensions to his advantage, will make any operator more efficient. In a more indirect way, mastering these basic facts helps an operator be better attuned to his machine and how it operates, so that his work flows more smoothly and safely.

The first step to becoming a good operator lies in learning the basic information reviewed here. As discussed in later chapters, the concepts of machine dimensions and weight distribution come into play in every aspect of TLB operation.

1-14

1-15
Swinging the load to the side, parallel to the TLB, makes transport easier (top), while filling the loader bucket helps stabilize the load being carried by the backhoe (above).

1-16 By swinging the hoe to one side, the TLB can maneuver in and out of surprisingly tight situations.

2. MACHINE CONTROL

General

Controlling the TLB means the operator knows how to make the machine perform exactly as he wants it to, regardless of any external conditions. In some situations, this may even involve forcing the machine to to what the operator wants, within the limits of safety. When an operator masters total control of his machine, common job site hazards and obstacles no longer prevent him from performing the work efficiently. Anything less will probably result in injuries or property damage, or both.

One common element that should be present in every TLB operation is **smoothness.** It is impossible to place too much emphasis on this basic concept. Smoothly operating the TLB should be the constant goal of every operator. Smooth operation will contribute to a longer machine life. It will also enhance the operator's efficiency; he'll be faster, and more precise.

Control comes into play during the two basic areas of machine operation, namely, **maneuvering the machine** and **hydraulic control.**

Maneuvering the Machine

Again, the emphasis must be placed on **control.** Anyone can get on a machine and drive it around a job site, but it takes a skilled operator to exercise total control over it. In the previous chapter we discussed several variable factors that affect the TLB's handling, such as weight distribution and how it can change during operation. A working knowledge of these factors, and how to manipulate them is essential to TLB control. The operator must develop a constant awareness of his machine's **total operating weight** (machine plus load), and how it is distributed.

Leaving aside the hydraulic features for a moment, the TLB essentially is nothing more than a tractor. The best way to maneuver the machine around a job site is to learn how to control its independent braking system. Each rear wheel is equipped with its own separate brake, with each brake controlled by a separate pedal. With the independent operation of the brakes, the TLB can be maneuvered in and out of surprisingly tight working situations. For example, by engaging one brake only, that wheel will lock while the other continues to drive. This will make the machine turn in a circle around the locked wheel. In many situations, the machine can actually be controlled better with the brakes than with the front wheels, which are actually designed to do the steering.

On slopes, learning to control the brakes will reduce the instability usually caused by weight transfer. As seen in Figs. 2-1 and 2-2, as the operator attempts to move the TLB up the sloped area, the front wheels leave the ground and control is lost as the machine veers off in the wrong direction. Figs. 2-3, 2-4 and 2-5 show the effect of applying the brake and forcing the machine to travel in the desired direction, with or without the aid of the front tires to assist in the steering. This control of the brakes is most usable on TLB models equipped with automatic, torque-converter transmissions. This is be-

2-1

2-2

2-3

2-4

2-5

cause the brakes can be applied on one wheel while applying any amount of power to the other wheel at the same time. On standard transmission models, the throttle is sometimes located near the brake and it becomes impossible to use both, which the operator sometimes must do in order to control the machine the way he wants. Obviously, this causes the brakes to wear more than they normally would. However, the extra wear is a small trade-off for the increased control and safety gained by mastering this technique.

There is also some loss of traction when traveling in reverse, again due to the tendency of the weight to be shifted up and off the rear wheels; but again, the backhoe attachment serves as an effective counterweight. Placing so much weight on the rear wheels, as noted earlier, can tend to make the machine too light on the front, particularly when operating on slopes. In this case, the front loader bucket should be filled, stabilizing the weight distribution.

When transporting material with the backhoe, which reduces weight on the front wheels, the brakes can be used for steering. Or, the operator can stabilize the front end by loading the loader bucket with heavy material, which will return steering control to the front.

Finally, one **fail-safe method** of maintaining control, especially in tight situations, is to simply **go slowly.**

Hydraulic Control

The TLB's hydraulic system controls the movements of the hoe itself, as well as the side-mounted stabilizers and the loader bucket in the front. Its heart is a pump and fluid reservoir located in the tractor. Without getting into a long essay on the principles of hydraulics, the basic idea is that fluid is pumped through the system at a flow ranging from fifteen to thirty-five gallons per minute. The operator controls the flow with six valves located at the rear of the tractor. These valves control the four basic hoe motions—the crowd, or "stick," the boom, the bucket and the swing—as well as the two stabilizers. The control for the front loader bucket is located near the steering wheel.

Depending on how they are manipulated, these valves send hydraulic fluid through the system to the function or functions the operator wants to actuate. As resistance is built up in the hydraulic cylinders, the force necessary to move the various parts of the machine is created. This force may reach several tons, particularly when the boom, stick and bucket are manipulated simultaneously to perform the most common backhoe operation—digging.

Since all power is controlled through these valves, the operator must become proficient in opening and closing them. This will determine how much force or speed is used to perform whatever task the operator faces. Each control is separate from the other, yet all are interrelated because they draw their power from the same pump. Therefore, if two or more valves are opened simultaneously, the oil will flow to the one with the least resistance. If all four main backhoe controls were actuated at once, the flow each would receive would be about one-fourth as much as each function would receive if it were operated alone. So, much of what it takes to control the backhoe depends on the correct diverting of hydraulic fluid from point to point. Loaders and other types of heavy equipment utilize their drive wheels to control and direct their force, but on the backhoe this force is controlled strictly through the hydraulic system.

Once the operator learns to operate the backhoe's hydraulics smoothly, and to go through the digging cycles with fluid-like precision, he has taken a big step toward **total backhoe control.** The smooth operator will bring the boom from the spoil pile to the trench and back again in sweeping, continuous arcs. Contrast this type of motion with the jerky, stop-and-go right angle movements of the beginning operator, and the benefits of practicing smoothness at all times are better appreciated.

To better illustrate the concept of smooth operation, we rigged a backhoe with a series of lights, and then took a number of time-exposures, while operating the backhoe (Figs. 2-6 and 2-7). The lights indicate the path of movement followed by the bucket, and give some idea of how the backhoe should be controlled, with the emphasis on smoothness.

It's impossible to learn how to be smooth from the pages of this or any other book or training manual. Control can only be mastered by making a concentrated effort to be smooth at all times while at the backhoe's controls.

The way to master smooth operation is through a technique called **feathering.** This is simply the opening and closing of the valves in combination, in such a way that the flowing style mentioned earlier is created. The beginning operator, using one function at a time, will find the machine rough, jerky and frustrating (Fig. 2-8, 2-9). High productivity will be impossible to achieve at first. Once he learns how to combine the hydaulic functions in a smooth way, however, he will be able to control the backhoe as he desires, and his productivity will rise to the level expected from professional operators. Again, this skill cannot be learned from a book, but only through **practice and constant striving for smoothness.**

One illustration of the use of feathering would be to swing the hoe across from left to right (or vice versa), carrying a load. As the bucket nears the end of the arc, start to extend the stick, which will then divert flow from the swing to the stick, and thus slow the swing. Then, at the end of the arc, roll the bucket outward, which means three functions are now in use. As the swing comes to a stop, bring the swing control back in the opposite direction; as the bucket reaches the end of its outward roll and the bucket lever is released, more power will come back to the swing and it will slowly, smoothly start back in the other direction.

A final note on hydraulics is that systems are becoming more sophisticated all the time. Recent developments include electro-hydraulic controls and limiting devices, which enhance precise operation. Also, an array of highly-engineered valve stacks further aid smoothness. Other examples of common valves found on TLBs include flow control valves and proportional priority flow divider valves; relief valves, and pressure reducing valves. All of these contribute to increased operator control, ease of operation and protection against hydraulic system damage.

In addition, some backhoe models now feature two-lever systems replacing the four-lever systems, combining two functions in each lever or "joystick." The advantages of the two-lever system include less operator fatigue and increased control when using all four functions at once, which is much more difficult on four-lever models. The four-lever system, on the other hand, gives the operator slighly greater backhoe control, especially when propelling or maneuvering the machine with the backhoe itself.

2-6
By following the path of the lights, viewer can trace the smooth, flowing digging cycle from right to left in the top photo, and from left to right in the photo below.

2-7

2-8

When contrasted to the photos on the facing page, these show clearly the jerky, inefficient digging motion common to beginning operators.

2-9

Practice Exercises

As mentioned earlier, machine control is not enhanced simply by reading this book. It is gained through practice, with an emphasis on smoothness. Here are some exercises that will help the beginning operator achieve smooth backhoe operation.

"Floating the Bucket." This exercise will teach the operator the most important basic motion of backhoe operation, the digging motion. When properly performed, the basic digging motion of the backhoe will move the bucket in long, level passes. The ability to perform this movement smoothly is essential to backhoe operation and is required in all precision excavations. How easy it is to perform depends on the hardness of the soil. If soil is hard, it will be fairly easy; if it is soft, muddy or sandy, it will become very difficult. A good operator must be able to handle all soil conditions, whether hard, soft or in-between. This exercise will teach the operator how to control the backhoe in the most difficult situations.

The object of this exercise is to follow the contour of the ground with the bucket teeth raised one inch above the surface. Extend the backhoe out all the way and lower it to within about an inch of the ground. Now, by raising the boom and pulling in the stick at the same time, the teeth will follow the contour of the ground while remaining just barely above it. When the bucket reaches the nearest position, reverse the procedure and extend the stick while lowering the boom. Practice this procedure in both directions until it becomes easy and speed is increased.

"Open area smoothness." To further help the beginner understand backhoe control basics, it is helpful to separate job functions into two categories. The **first category** includes jobs which don't require precision digging or spoil placement. These may include any task in an open area with plenty of swing room for operating the boom, and without height or exact spoil placement restrictions; any time the operator practices in an open area; and when digging large excavations.

In these situations the operator is not required to maintain precise machine control, because there is nothing within the backhoe's reach that can be damaged. The machine can be operated at maximum speed and efficiency. These are excellent opportunities for the operator to practice and concentrate on developing a perfectly flowing, smooth digging cycle (Fig. 2-6). The beginner can work on mastering the art of feathering the controls, enhancing smoothness. The best thing about this kind of practice is that it can be done on the job, as long as it is safe. Because there are no obstacles or restrictions, this general digging motion is ***easiest*** for the beginning operator to perform—yet in reality, it is not at all easy, and usually takes hundreds of hours to become accustomed to. When it gradually comes naturally to the operator, however, he can then begin concentrating on the more difficult tasks.

The **second category** of basic backhoe functions— the more difficult jobs—usually make up about 80 percent of all tasks performed by the TLB. These jobs include obstacles and restrictions which demand a high degree of control, to prevent damage to the area or structures in the area. These restrictions can include: houses and buildings; lawns, trees or bushes; delicate concrete, asphalt or masonry surfaces; trenches or other excavations; existing footings; building materials piled on the site; cars and trucks; and stakes, hubs or other surveyors' markings.

Another set of obstacles exist underground. Transportation systems for gas, water, sewage, electricity, telephone wires, and cable television can pose problems, as can oil, steam, and cold and hot water lines. On some job sites, everything from gases and chemicals to beer or grape juice can be piped underground.

Some jobs may include only a few of these restrictions, while others are so complicated that safe, damage-free TLB operation can be next to impossible. However, it is the operator's job to control the machine in such a manner that damage will be prevented. It is also his responsibility to know when job site conditions are so restrictive that damage could be unavoidable. In these cases, the operator must inform those in charge of the job, or, as a last resort, he should know when a job shouldn't even be attempted.

To help the beginning operator gain the type of control necessary in these difficult situations, the following exercise may be practiced.

"Dipping Water." To practice this exercise, some source of water will be needed. This could include a stream, pond or similar source that will allow the operator to move the water about without causing problems, and one which allows easy, safe access to the TLB.

The object of this exercise is simple: dip the water with the backhoe bucket, swing it to the side and dump it, simulating the act of digging. The beginner (and even many experienced operators) will have a hard time avoiding splashing or spilling the water at first. As smoothness and control increase, however, the operator will be able to keep the "load" in the bucket, and avoid splashing or spilling.

Unlike the "open area" exercise where the operator can work at full speed without worrying about restrictions, this exercise simulates working in extremely tight situations, where one quick or thoughtless move could cause damage or other accidents. It simulates the kind of situation where every move must be planned and executed carefully, with the utmost in precision. By practicing swinging in both directions, this exercise is an excellent way to gain the control necessary for most on-the-job situations.

Practicing these exercises and becoming gradually accustomed to smooth, controlled TLB operation while off the job will enable the beginning operator to develop the confidence necessary to perform the duties expected of him and his machine while on the job.

3. SAFETY

General

No discussion of the operation of any piece of heavy equipment is possible without a strong emphasis on safety. Every move made with the TLB must be done so with safety in mind—safety to you, safety for those around you, safety to property, and safety to the TLB itself. Indeed, safety considerations must be made even before getting on the machine. According to the State of California Department of Industrial Relations, the leading cause of non-fatal disabling work injuries involving backhoes in 1980 was simply slipping, falling or being struck while getting on or off the machine.

It is also quite possible for someone on the job site to be injured due to short-sighted TLB operation, and yet never come in direct contact with the machine. The careful operator must be aware of all possible dangers, at all times, and do everything he can to reduce the risk of injury or accident. Anything less could cause injury or even death, to the operator or the workers around him, who are also his responsibility. **The operator is always responsible for the safe operation of the TLB.**

Every part of this guide, regardless of the technique being discussed or the information being given, has some aspect of safety underlying it. There are any number of possible hazards to beware; literally anything could go wrong at any time. Usually, though, the potential for mishap can be greatly minimized if the operator makes good, safe operation of the TLB his prime consideration at all times.

Many accidents involving construction equipment occur because the operator or helpers fail to follow some very basic rules. When operating equipment, just remember that a little forethought and common sense can prevent some very serious injuries to you or someone else. Take just a moment to consider these basic rules:

- NEVER TAKE ANYTHING FOR GRANTED. Don't assume that everything is all right at the start of the work day just because everything seemed all right at the end of work yesterday. Before beginning operation, thoroughly inspect the equipment to be sure it is in good operating condition.

- KEEP THE MACHINE CLEAN, including all windows, glass and light lenses. Remove any excess oil and grease. Store tools and other necessary items in the tool box.

- CLEAN MUD OR GREASE from your shoes before attempting to mount or operate the equipment. Be sure the foot pedals are clean and dry to reduce the possibility of your shoes slipping off the pedals. Also make sure to keep all hand-operated controls clean and free of grease.

- CLOTHING should be relatively close fitting. Loose jackets, shirt sleeves, rings, and other jewelry should be avoided because of the danger of catching them in moving parts or on controls.

- ALWAYS WEAR required protective items such as hard hats, safety glasses, reflective clothing, safety shoes and ear protection.

- NEVER START OR OPERATE without protective guards and panels in place.

- ALWAYS CHECK height, width, and weight restrictions for the area and be sure the equipment will not exceed those limitations.

SELECTED OSHA CONSTRUCTION INDUSTRY GUIDELINES

(Courtesy State of California Department of Industrial Relations)

EXCAVATIONS, TRENCHES, and SHAFTS

A permit issued by DOSH is required prior to starting work on excavations, trenches, and shafts which are 5 feet or more deep and into which employees are to enter. 8 CAC 341

Prior to excavating, the location of underground utilities must be determined and utility owners must be advised of the proposed work. 1540(a)(1)

EXCAVATIONS

Before any employee may work in or adjacent to the excavation, the employer must inspect any excavation for hazards from possible moving ground. 1540(a)

Excavations must be inspected by a qualified person after every rainstorm, earthquake, or other hazard. 1540(a)(3)

Excavation Shoring System

Shoring must be properly designed and consist of sound wood timbers, and sheathing or sheet piling as needed. 1540(c)

The sides or walls of an excavation may be sloped at ¾:1 (horizontal:vertical). Where the soil is unstable, a greater slope is required. 1540(d)

Sloping may be combined with shoring.

Benching

The sides of excavations or faces may be guarded by cutting back ½ the face. 1544(b)

Excavation Safety Rules

- There must be proper, qualified supervision at all times during excavation. 1540(g)
- Safety provisions must be taken to protect workers while installing or removing shoring systems. 1540(h)
- Special safety provisions must be taken to protect workers in areas subjected to vibration or extra loads from heavy equipment. 1540(j)
- Spoil must be kept back at least 2 feet from the edge of all excavations. 1540(f)
- Safe and convenient access to excavations must be provided. 1540(i)
- Effective barriers must be installed at excavations adjacent to mobile equipment. 1540(n)
- Water must not be allowed to accumulate in any excavation. 1540(m)

TRENCHES

Trenches are excavations in which the average depth exceeds the width, and the width is 15 feet or less at the bottom. Shafts, tunnels, or mine excavations are not trenches. 1540

All trenches 5 feet or more deep must be guarded against the hazard of possible moving ground. Trenches less than 5 feet deep must also be guarded when this hazard exists. 1541(a)

Trenching Safety Rules:

- Before entry into a confined space, the air must be tested for dangerous contamination and/or oxygen deficiency. 5158(c)
- Exit must be provided at 25 foot intervals for all occupied trenches 4 or more feet deep. 1541(b)(2)
- Safe crossovers must be provided when needed. 1541(c)
- Safety provisions must be taken to protect workers engaged in installing shoring. 1541(b)(1)
- Covers placed over open trenches in roadways must be secured against displacement.

Large excavations must be inspected daily. 1546(a)

Excavation guarding

The walls and face of all excavations 5 feet or more deep, into which employees will enter, must be guarded by a shoring system, sloping the ground, or other equivalent means. 1540(b)

Alternative shoring or sloping systems may be used only when designed by a civil engineer licensed in California. 1540(c)(4)

HEAVY CONSTRUCTION EQUIPMENT

GENERAL

- Repairs must not be made to powered equipment until workers are protected from movement of the equipment or its parts. 1595(a)
- Wherever mobile equipment operation encroaches upon a public thoroughfare, a system of traffic controls must be used. 1598(a)
- Flagmen (wearing high-visibility vests) are required at all locations where barricades and warning signs cannot control the moving traffic. 1599(a)
- Jobsite vehicles must meet design requirements as follows:
  - Operable service, emergency, and parking brake. 1597(a)
  - Two operable headlights and taillights for night operation. 1597(b)
  - Windshield wipers and defogging equipment as required. 1597(c)
  - Seatbelts are required if the vehicle has Roll-Over Protective Structures (ROPS). 1597(f)
  - Fenders or mud flaps are required. 1597(g)
- Vehicles used to transport employees must have adequate seating. 1597(e)
- Vehicles (and systems) must be checked for proper operation at the beginning of each shift. 1597(h)

- BE SURE that all safety devices provided are in place and in good operating condition. Be familiar with the use of all safety devices, including lights, seat belts and back-up horn.

- PLAN AHEAD—WORK SAFELY—Avoid accidental damage and injury. If an accident does occur, react quickly with the tools and skills at hand—know how to use a first aid kit and a fire extinguisher, **and where to get assistance.**

- LEARN—BEFOREHAND—as much about the working area as possible. Be sure that exact locations of obstructions or hazards are known. Such locations should be precisely marked by the operator to prevent accidents.

Each year there are hundreds of fatalities and thousands of serious, lost-time injuries as a result of TLB and other heavy equipment accidents. Most of these could have been prevented, if the operator had only used better judgment and common sense while on the job. Heavy equipment works hard and makes many people's jobs easier, but it must be treated with respect—before, during and after the operations.

The condition of the TLB must be checked **before** starting work. Items to check include the steering, the brakes, the clutch, the hydraulic system and the electrical system. Any defect should be reported at once. Never start work with a defect that could cause an accident. Otherwise, the unsafe condition could be compounded by the operator's unsafe act.

The safety canopy, or Rollover Protective Structure (ROPS), was designed to protect the operator from falling materials, as well as from rollovers. Make sure the ROPS is maintained in good condition and is adequate for the type of work being performed.

Remember to use the seat belt **at all times.** On a rough grade, the belt can keep the operator in his seat, which means he will maintain control of the TLB. If the TLB turns over, the seat belt will keep the operator from being thrown. The ROPS is designed to prevent rollover injuries, but the operator must be buckled in during the roll.

Before we move into a discussion of the major types of TLB danger, it is sufficient to say **there is no substitute for common sense.** Simple things like going slow when unsure of ground conditions, or learning to always watch out for other workers because they may not be watching out for themselves—these sound pretty basic. When the operator takes them for granted, and lets his concentration lapse, however, trouble begins. Never forget that when the operator is at the controls of the TLB, the burden of safety is on him.

The most obvious danger presented by TLB operation is the sheer size and bulk of the machine. In addition to the actual machine weight of approximately seven tons, the TLB also generates tremendous hydraulic power. This kind of weight and power can cause crushing injuries. Practically any type of injury is possible, ranging from something as minor as a cut or scrape to the major accidents that can cause permanent disability or even death.

Backhoe Accidents

According to the Department of Industrial Relations of the State of California, non-fatal disabling work injuries involving backhoes in 1980 were classified as follows:

Accident type	**number of injuries**
Fall or striking backhoe while climbing on or off	16
Struck by backhoe bucket	14
Struck against backhoe (not classified elsewhere)	14
Caught in or between backhoe parts	14
Overexertion involving backhoes	10
Backhoe overturned	6
Other	10
TOTAL	84

These accidents can take many forms depending on the conditions and circumstances surrounding each. Here are some examples (all taken from actual accident reports):

- An equipment operator was working in an excavation when the backhoe's stabilizer lodged in the trench wall. The backhoe overturned, and the operator's back was injured.

- A construction worker was releasing the latch on a backhoe bucket when the operator inadvertently started the equipment, catching the worker's hand and breaking two bones.

- A maintenance worker received a hernia when his foot slipped on the wet step of a dew-covered TLB.

- A worker was shoveling dirt next to a TLB in operation. The bucket overswung the trench and struck the worker, resulting in a broken leg and ankle.

- An operating engineer was leaning over the edge of a trench between two shoring plates, measuring the depth of the trench. A backhoe hit one of the shoring plates, pushing it into the other with a scissoring effect on the engineer's neck.

- A water utility worker was helping a TLB operator change a bucket. While aligning the holes of the pin mechanism, he jammed a thumb and index finger.

- A grading and paving operator was working on a hillside, trying to remove a stump, when his machine rolled over. His ankle was broken when the roll bar hit it.

- A worker pulled a back muscle while trying to move a 24-inch bucket.

- An operator struck his head on the bucket when he slipped getting off the machine.

Loader Accidents

California's Department of Industrial Relations also tabulates injuries received while operating front-end loaders, which are relevant to TLB operations. A breakdown of statistics for 1980 details the types of front-end loader accidents.

Accident type	number of injuries
Struck against loader (not classified elsewhere)	42
Caught in or between loader parts	24
Falling or striking against while climbing on or off	20
Struck by bucket	16
Operating loader over rough or uneven terrain	14
Overexertion involving loader (not classified elsewhere)	20
Loader ran over embankment or overturned	6
Run over by loader	2
Other	10
TOTAL	154

Again, some examples from actual accident reports show the types of mishaps that can occur:

- A laborer was in a trench attaching a shoring pulling device. A second worker, relaying hand signals to the loader operator, failed to halt the lowering of the front bucket. It stuck the first employee on the head, causing spasms, stiffness and soreness in his neck and back.

- An equipment rental operator accidentally drove a front loader in a trench, causing the loader to become off-balance and vibrate violently. The operator was bounced around inside the cage, hit his head on the beam, and suffered head injuries.

- A job foreman hit a rock with the loader. His head hit and broke the windshield, resulting in head and neck soreness.

- An operator who was breaking and removing concrete backed over some rocks, came down hard in his seat, and experienced sharp back pains.

- An operator had overloaded the front-end bucket on his loader, causing the machine to tip forward. As it did so, the operator grabbed a metal bar on the front of the cage. As the loader returned to a level position, the bucket came back, catching his hand between the bucket and the metal bar. His hand was "severely lacerated," according to reports.

- A truck driver was unloading a front-end loader from the back of a transport trailer. Because he backed the machine in a curved, rather than straight line to the back of the trailer, it slipped off the trailer and came to rest on its side on the ground. The driver was crushed by the canopy part of the ROPS. He had been thrown from the driver's seat on the loader because he wasn't using his seat belt.

- An operating engineer was asked to drive a loader to a construction site. He drove the loader in high gear along a 6% grade road. He lost control, hit the shoulder of the road and rolled over. The gas tank exploded into flames and the operator was killed.

- A survey team was laying out and grading an area prior to concrete work. A loader was being used to level the land. When the loader made a stop, an operating engineer got behind the loader to drive a grade stake into the ground. The loader operator did not notice him, and began backing up. The equipment (which lacked a back-up horn) ran over the engineer, killing him.

Rollover

The unique weight distribution characteristics of the TLB make safe operation particularly difficult, because they create many hazards . . . some of which the operator cannot clearly see at all times. Because of the weight and the sometimes awkward way it is distributed, one of the most common hazards to the operator, as we have seen, is rollover. It is not unusual for the operator or those nearby to be severely injured or killed in such accidents. Even with the introduction and requirement of ROPS on all TLB's, injuries still occur when the ROPS can't withstand the entire weight of the machine being thrown on it, or when the operator neglects to wear his seat belt.

Between 1960 and 1975, 229 California workers were killed when the equipment they were operating rolled over or ran off the road. It is useful to examine a breakdown of the types of industries involved in these accidents:

Type of worker	number of deaths
Construction	139
Agriculture	20
Mining	19
Logging	18
Government (street maintenance, public utilities)	15
Equipment rental	2
Other	46
TOTAL	229

Let's look at a breakdown of the types of accidents recorded in California during the period 1966-1975:

Accident type	number of deaths
Operating too close to the edge of road or embankment	40
Operating on steep or uneven ground	20
Equipment defective or malfunctioned	11
Operating equipment on uncompacted or otherwise unstable ground	10
Operating equipment at excessive speed	8
Loading or unloading equipment onto/off of flatbed truck	6
Equipment struck rock, brush, etc.	6
Ran equipment up enbankment	5
Operator lost control of equipment while turning	2
Other, or insufficient information to classify	21
TOTAL	129

Note that in the above table, almost one-third (40 out of 129) occurred when the operator moved his equipment too close to the edge of the road or embankment.

Here are some specific accident descriptions reported to the state:

- A truck driver was using a loader equipped with a rear-mounted drag, pulling some form panels up over a fifteen-foot high bank. The panels had been stripped from a retaining wall and were lying on the slope that had been created when the area was excavated for the retaining wall. The tractor slid backwards and went over the edge of the bank. The rear wheels became wedged between the retaining wall and the dirt bank, and the front of the tractor flipped upward into a vertical position. The operator was pinned between the tractor and the wall. It was later learned that he was not qualified to operate the loader in the first place.

- An operating engineer was running a front-end loader on a rural road about twenty-five feet wide. On the uphill side of the road, a trench was being dug for the installation of a storm drain. The downhill slope from the road was wooded, and sloped about 45 degrees. The operator planned to go for a load of backfill material, and proceeded to turn the loader around. As he did so, one of the rear wheels went over the edge of the downhill bank, and he lost control of the rig. The equipment began rolling, and the operator was crushed beneath the machine about thirty feet down the bank. The machine kept rolling until it came to rest about 200 feet below the embankment.

These are only two of the many ways an operator can be injured or killed by rollover accidents. The risk of rollover can be reduced by monitoring how the machine's weight is distributed and transferred while work proceeds. Essentially, the total weight of any piece of heavy equipment is distributed down through the ground at a 45-degree angle in all directions. It is vital, when preparing to operate the TLB, to constantly consider **how its weight is going to be supported.** Try to visualize the work area as a "platform" on which the TLB will rest. Once this working platform has been established with the stabilizers, wheels, or loader bucket all firmly supported, the operator must constantly think about how the machine's weight distribution will change as work progresses. He must consider not only the machine's weight, but the weight of the material being handled as well.

Soil Conditions. The composition, and therefore the stability, of the soil in which the TLB is working can be crucial to causing rollover. The operator must investigate whether the soil has a high water content, or if it contains mostly sand, or gravel, or some other unstable mixture. Even stable-**looking** soil could be hazardous—it could conceal an old, covered trench that could lead to a cave-in, rollover, or both.

Excavations

Excavation work is considered among the most hazardous of all construction operations—and the TLB is primarily an excavating machine. Unless precautions such as shoring or sloping the trench walls are taken, cave-ins can occur; if this happens, workers in or near the trench could be injured or buried, and equipment near the trench could topple over into it.

As is the case with rollover, there are plenty of grim statistics which make the dangers of cave-ins clear. In the ten years from 1964-1973 in California, for example, eighty workers were killed in cave-ins. The reasons for these deaths were given as follows:

Reason for cave-in	fatalities
No shoring or sloping	31
Shoring installed/removed in unsafe manner	15
Inadequate shoring or sloping	11
Working outside shored area	9
Shoring was just removed	2
Weight of adjacent pile or wall	2
Other (or not stated)	10
TOTAL	80

Despite laws requiring shoring or sloping, you can see that 57 out of 80 cave-in deaths (about 74 percent) were caused by **inadequate** shoring or sloping, **improperly installed** or removed shoring, or **no shoring or sloping at all.** The law requires that shoring work start at the top of the trench or excavation and work downward, so that workers are protected by the cross-braces already in place. The procedure is reversed for removal. **At no time** should a worker be inside a trench pulling out a jack above him.

Cave-in deaths occur primarily due to suffocation or crushing. Of the eighty workers killed in the period noted above, 47 suffocated when they were buried, while 26 were crushed by falling debris. Six sustained fatal skull fractures, and one worker died of a neck wound caused by a sharp object.

Specific examples of cave-in deaths, and the ways they were caused, include the following:

- A TLB was excavating a twelve-foot trench for installation of a storm drain. A laborer was standing on the bucket of the backhoe, placing the lower screw-jacks against the uprights, which were spaced about three feet apart. The trench was in an area that had been previously filled. The bottom half of the trench suddenly caved in, knocking the laborer off the bucket and covering him with five feet of dirt and sand. Another worker was pushed against the opposite wall of the trench by the cave-in, but was uninjured. The first worker suffocated.

- Two laborers had been told to compact the backfill in a fifteen-foot trench and remove the shoring. One man, who was standing on about seven feet of backfilled soil in the trench, removed a lower jack which was bracing the uprights, and then tried to remove the upper jack. When it was removed, the trench wall collapsed and he was crushed—a victim of improperly removed shoring.

- A sewer line project had been completed, but leaks were discovered when the line was tested, so the process of re-excavating and patching the pipe began. While this happened, water began accumulating in the bottom of the ten-foot deep trench. Workmen decided to dig a perpendicular trench to provide drainage. A laborer who had been in the excavation looking for leaks stood back in a section of the original trench, as the TLB began to dig the second trench. Almost immediately a section of the wall adjacent to the operation slid in and trapped the worker, who was in a stooped-over position. Because the face of the excavation was very irregular, it would have been difficult to install efficient shoring without dressing down the cut face. In this situation, the sides should have been properly sloped.

Excavation Safety

Obviously, one of the keys to preventing cave-ins is, again, knowing the **ground conditions.** Ground that appears stable may not be due to previous excavations. Or it may be loose, or muddy, or contain sand and gravel. In order to reduce the risk of cave-ins, the Occupational Safety and Health Administration (OSHA) now requires that all trenches deeper than five feet either be shored or else sloped at a ratio of three-quarters of a foot across in both directions for every foot in depth. While shoring is generally not the responsibility of the TLB operator, sloping the trench is done with the TLB. (See **Sloping.**)

Another point to consider is, when digging an excavation, do not place spoil closer than two feet from the edges, as the weight can cause a cave-in.

The operator must also take care to **never** undermine existing footings, such as piers or walls. Removing the support from under these or similar heavy structures could cause them to collapse either onto the TLB and surrounding equipment and workers, or they could collapse later onto unsuspecting workers at the job site.

In some situations, the TLB operator may be the only person on the job site who is able to evaluate things like existing footings or possibly unstable ground conditions. He must learn to watch for telltale signs of a cave-in about to occur, because he may be the only person in position to see them. Soil may begin to fall from the sides of a trench, either up near the edge or down toward the floor. Cracks may appear in the earth around the trench edges, either parallel with the excavation, or at an angle to it. Any of these signs indicate that movement is taking place in the ground around the trench, and it is the TLB operator's responsibility to **act quickly,** warning workers in the trench to get out **as fast as they can.**

If his machine is in a position where it might fall over into the excavation, the operator may have only a few seconds to either get the TLB clear of the area, or if that is impossible, to get off the machine and reach safe ground. If unstable ground conditions are present, take care not to cause the TLB to shake or move in a jerky manner, since this may precipitate a cave-in, especially if the excavation is deep.

Excavation Safety Q & A

Test yourself on the following excavation safety-related questions. The answers can be found at the end of the chapter.

1. According to OSHA, when must a trench or excavation be supported by shoring, sheeting, bracing or sloping?
2. What is the minimum distance excavated material may be piled from the edge of the excavation?
3. What is the maximum distance a worker should have to travel to reach a ladder leading out of the trench?
4. What is the "angle of repose"?
5. Do sources of vibration such as nearby vehicles, heavy equipment, railway traffic, blasting or pile drivers materially affect soil stability?
6. When the slope of the excavation approximates the angle of the excavated material, is it reasonable to assume the excavation is a safe work area?
7. The usual compliance time to correct an unsafe excavation condition is: A. one day B. two days C. five days D. at once
8. Equipment working near high voltage electric power lines shall have a clearance from the point of operation to the lines of at least: A. six feet B. fifteen feet C. twenty feet D. ten feet
9. What is the most common cause of trench and excavation cave-ins?
10. What is the only safe procedure to follow when installing or removing shoring systems from trenches or excavations?

Shoring

As we have noted, the vast majority of trench cave-ins are caused by insufficient or improperly-installed shoring. Remember also that shoring is usually not the responsibility of the TLB operator, although he is often called upon to assist the workers who actually install the shoring. A quick overview of the conditions which affect the kind and amount of shoring necessary includes the following points, which the TLB operator should always keep in mind:

- **The depth of the trench.** Remember that any trench five feet deep or more must always be shored or sloped. If there is any possibility of soil movement, even shallower trenches should be shored. If there is any doubt, leave nothing to chance—shore or slope the trench.
- **Running soils.** The looser, or more liquid the soil, the greater the need for precautions against cave-in.
- **Changing weather conditions.** Hard-packed soil can become muddy and unstable after rain.

Trenches which had been safely shored or sloped in dry weather can become extremely dangerous when the ground is wet. Thawing soil can also become unstable quickly.

- **Heavy loads in the area.** Don't park heavy equipment next to a trench. Don't stockpile any material closer than two feet from a trench edge. Also, nearby objects such as buildings, curbs, trees and utility poles will exert stress on trench shoring.
- **Vibrations.** If a trench is being dug near a roadway or where other operations create vibrations, make the shoring strong enough to withstand the added stress.

Sloping

Unlike shoring, which is usually accomplished by other workers, sloping is a process which the TLB operator is usually responsible for. For this reason, it is important for the operator to thoroughly understand both the procedure as well as the desired results. Sloping an excavation is much more than just digging a hole in the ground—the purpose is to provide a safe working area for the workers in the trench.

As noted earlier, when a trench is to be sloped, the required ratio is three-quarters of a foot in both directions for every foot in depth. Thus, the sloped area for a twelve-foot trench would measure twenty feet across (nine feet on either side plus the actual two-foot trench width). As Fig. 3-1 clearly illustrates, sloping requires removing a large quantity of material to meet these requirements. Sloping requires the removal of many times the amount of material that would be removed if the trench was simply shored. The rule of thumb to approximate spoil removal is that sloping will require the removal of material about equal to the amount removed by shoring, multiplied by half the depth of the trench.

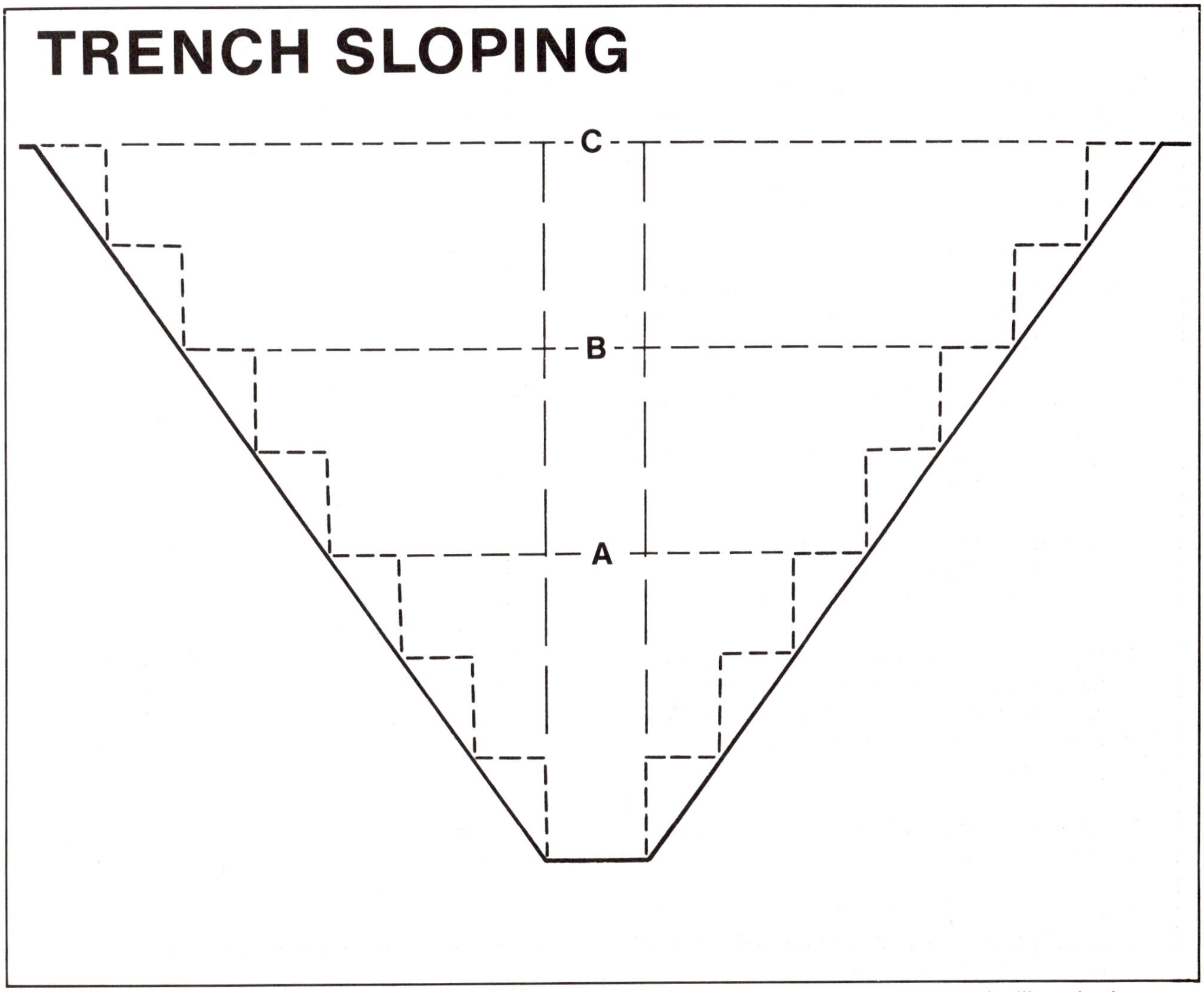

3-1. A. A six foot deep trench will require the removal of three times as much material. B. A ten foot deep trench will require the removal of five times as much material. C. A twelve foot deep trench will require the removal of about six times as much material.

3-2. A sloped trench is dug in steps . . .

3-3. . . . until the desired slope is accomplished.

3-4. The sides are then smoothed by sliding the bucket up and down against the slope.

3-5. The accumulated spoil at the top of the excavation is cleaned away before the final grading is done on the bottom.

It is obvious that the amount of spoil to be removed in the process of sloping becomes enormous as trench depth increases. This affects TLB work two ways: 1) Digging time is increased dramatically. 2) Spoil placement quickly becomes a problem and there is often not enough room for stockpiling in the immediate area.

Using a larger bucket will reduce digging time, but the operator must also carefully "dress" or smooth the sides of the slope, which is somewhat time-consuming (see Figs. 3-2 through 3-5). To help the TLB operator stockpile the material, a loader or bulldozer can be used to push spoil back, allowing the digging to continue. If a second machine isn't available, the TLB operator may have to frequently stop digging and move the spoil back from the excavation using the loader.

Transporting Wheeled Machinery

The first major point made in this chapter was that the size and weight of the TLB and similar machines alone make their operation inherently dangerous. This extends to the transporting of machinery to and from the job site, and its loading or off-loading from transport trailers or other transporting vehicles. Careless handling of such heavy equipment when transporting it can prove just as deadly as careless operation. Here are some suggestions to keep in mind which can help reduce the risk of accidents during heavy equipment transport.

Prior to loading:

- Be thoroughly familiar with all basic machine controls—levers, pedals, switches, gauges, etc. Test drive it to familarize yourself with the controls as well as the machine's performance; get the "feel" of how it responds. Unless absolutely sure you know how to operate it, don't—get someone else to do it.
- Check the operator's manual to review physical dimensions of the machine. Be sure its width complies with state highway standards, and that proper permits have been obtained. Check rules regarding use of warning signs, flags, escorts, etc. Make sure that the machine's weight, as specified in the operator's manual, can be supported by the transporting vehicle.
- Place chocks or other blocks in front of the transporting vehicle's wheels to prevent it from rolling while loading or unloading.

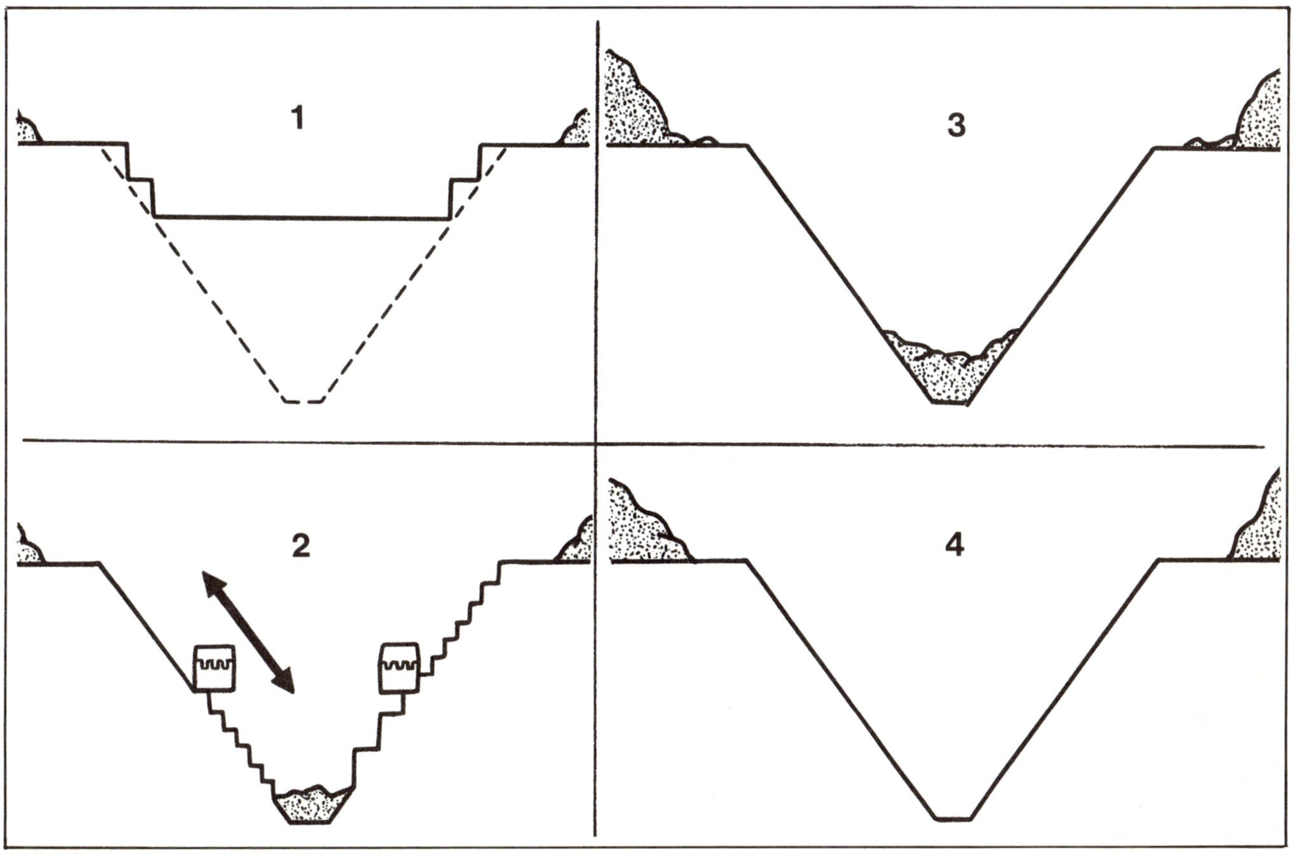

3-6. SLOPING PROCEDURE: 1) After the correct width is determined, digging proceeds at the outside edge of the excavation. 2) After bulk material removal, slope sides are notched into smaller notches, and then the bucket is used to grade the sides smooth. 3) After the slope has been graded, the "sidewalks" are cleaned and the bottom of the trench is fine-graded. 4) A finished section.

- Make sure there are no slippery surfaces on the bed of the transporter. If the bed is wet or covered with some substance that will reduce traction, dry it or clean it.

During loading:

- Load the machine from level ground. The ramp should be strong, secured, and long enough to provide a slope of low angle. There should be sufficient height for proper entrance onto the transporting vehicle.
- Make sure that buckets, blades, forks and other mounted equipment are lowered, not obscuring the transporting vehicle driver's vision, making the load too top-heavy, or striking surfaces or obstacles while loading.
- Approach the ramp and transporting vehicle in the direction and manner recommended in the operator's manual.

Following loading:

- Lower and secure buckets, booms forks and other mounted equipment onto the bed of the transporting vehicle.
- Shut off the engine, and remove the key from the ignition.
- Place the transmission in "park" or low gear. Apply the parking brake.
- Secure chocks before and behind the wheels or tracks of the TLB or other equipment being loaded.
- Secure the TLB to the transporting vehicle with chains or cables. Usually, suggested tie-down points are identified in the operator's manual. If none are suggested, common tie-down points include the axle housing, openings in wheels, front axle members, tow hooks, and drawbar mounts.
- Check to make sure the tie-downs are not in contact with the machine's equipment hoses, hydraulic cylinders, valves, rods or tires.

Conclusions

In general, accidents occur when the operator either: 1) fails to recognize a potentially dangerous situation quickly enough to prevent it; or 2) he recognizes danger in time, but fails to investigate possible solutions, or to consider possible consequences.

This failure to act is probably due to the fact that out of every 330 accidents, only one is a major mishap involving serious injury or death, and only twenty-nine require any medical attention at all. The remaining 300 accidents are classified as "minor"—the ones that are easy to forget about. Unfortunately, TLB operators are subject to the law of averages just like any other occupation or random group of people, and those many, many "minor" accidents have a way of catching up.

The TLB operator must be sure of every move he makes. Whenever in doubt, he should stop what he is doing. If he is unsure of the safety of any operation he has been asked to perform, he should question it. If a certain maneuver or job seems too dangerous, it would be better for the operator to simply refuse to do it, rather than risk lives and property.

If danger ever develops in the middle of a job, stay calm and think yourself through the situation. The earlier danger is recognized, the easier it can be avoided. Many operators don't recognize danger until it is too late, as we have seen. So think ahead, and come up with a plan. Consider all the possibilities. If the TLB becomes stuck, for example, don't sit there and spin the wheels; get off, walk around the vehicle, and think: Where is the weight? How is it distributed? How **should** it be distributed? Come up with a well thought-out plan to get from where you are to where you want to be.

Look out for workers and other people on the job site who are too ignorant or careless to look out for themselves. Make sure no one wanders into the swing range of the machine, and if they do, tell them to move to a safe distance.

If you suspect there are hidden dangers such as unstable ground conditions, speak up. An electrician, or a plumber or property owner probably don't know as much about safe TLB operation as the operator does. Don't be afraid to voice an opinion on what procedures should be followed.

Use your head, consider all possibilities, observe common sense, and be safe every minute the TLB is in operation. A safe operator is the best defense against needless TLB accidents.

ANSWERS TO EXCAVATION SAFETY EQUIPMENT QUESTIONS ON PAGE 20.

1. *When the trench is five feet deep or more, or when soil is particularly unstable, or when men will be working with their heads below the surface level.*
2. *Two feet. In addition, objects like pipes, rounded boulders, etc. should be secured to prevent them from rolling into the trench.*
3. *Twenty-five feet. The ladder should extend three feet above the ground surface level.*
4. *It is the greatest angle above the horizontal plane at which excavated material will lay without sliding.*
5. *Yes. Vibrations are a frequent cause of cave-ins.*
6. *Yes, generally speaking.*
7. *D - at once.*
8. *D - ten feet.*
9. *The most common cause of cave-ins is inadequate or improperly installed shoring, usually in an attempt to save time or cut costs.*
10. *When installing shoring work from the top down, closely following the digging. When removing shoring, work from bottom up. Jacks or braces should be removed slowly, and ropes should be used to pull out the jacks and braces from above.*

4. TLB MAINTENANCE

Like any piece of precision equipment the TLB requires maintenance on a daily as well as periodic basis. Most of the daily checks are (or should be) performed by the operator, since he is familiar with the machine's performance day to day. Besides the obvious benefit of fewer problems and less downtime on the job site, regular maintenance will enhance the TLB's long-term performance and prolong the machine's life, giving its owner the most for his investment.

Daily Maintenance

The most important daily maintenance the operator performs is to check all of the machine's fluid levels and replenish them as needed. These include engine oil; power steering fluid; transmission fluid; hydraulic fluid; engine coolant; and diesel fuel.

Besides knowing his fluid levels at all times, the TLB operator also needs to know what grades or types of fluids to use in his particular machine. Fluid specifications vary from manufacturer to manufacturer, so it is wise to follow your machine manufacturer's recommendations.

Greasing the TLB daily is also very important. There are numerous mechanical joints located throughout the machine, and all of them have grease fittings.

Fluid levels should be checked prior to the warm-up period. Once the engine is on, the operator should check the various gauges on the instrument panel to make sure oil pressure, engine temperature, etc. are rising to normal operating levels. If a gauge shows an unusual condition such as low oil pressure or high engine temperature, do not operate the machine until the problem has been corrected.

While working, the operator should also be alert for changes in the operation of the TLB's hydraulic system. If the pump makes an unusual noise or whine, or if slow or jerky operation of the hydraulic functions is noted, it probably means the fluid level has fallen too low, allowing air to work into the system. Operating the machine with insufficient hydraulic fluid could damage the pump and the system itself.

Warm-Up and Cool-Down

Because of the nature of diesel engines, they require a period of warming up in the morning (or before being used under load), as well as a brief cool-down before shutting the engine off. Along with keeping the oil and other fluid levels maintained, this has the greatest effect on prolonging machine life. Each manufacturer has its own recommended warm-up procedure; some recommend idling the engine, while others suggest running at half-throttle. A good rule to follow, when in doubt, is to run the engine just slightly above idle for at least five minutes.

The warm-up period is necessary to bring the engine and hydraulic system up to running temperature; therefore, cold weather starts require somewhat longer warm-up time. The warm-up also turns the gears in the hydraulic pump, warming the fluid and circulating it through the system. Once the machine has warmed up, the operator should check to make sure the various hydraulic functions perform smoothly.

While the engine is warming up, the operator should get in the habit of looking over the machine, quickly checking such points as tire pressure, hoses and connections (look for leaks or signs of wearing), and the bucket pins and clips, to make sure they are properly seated. This is also a good time to get in the habit of greasing the machine regularly.

Machine Life

When one looks at the TLB itself, and considers the tremendous stress and force associated with its operation, and the number of hinge joints that the hydraulic power is transmitted through, it becomes clear why preventing wear in these joints is the most important aspect of extending the TLB's machine life.

The two most important factors which contribute to unusual wear are:

1. Unnecessary stresses, strains or shocks to the structural bearing surfaces. To prevent such stress and strain, the TLB must be operated in as smooth and flowing a style as possible (See chapter on ***Machine Control***.)

2. Failing to grease the bearing surfaces on a regular basis. Without this lubrication there is nothing to keep these surfaces from wearing quite rapidly. Once they become worn, the resulting looseness will cause the surfaces to hammer against each other, causing even more wear. It should also be noted that the hinge joints do not have any seals to keep grease in or dirt out, so they must be greased especially frequently.

Greasing Tips

When greasing the wear parts on the TLB, some helpful suggestions include the following:

- Make it a habit of greasing the machine lightly and frequently, as opposed to a heavy greasing less often. This is the best protection against wear, and it also prevents dirt from collecting in and clogging the grease flow passages (Fig. 4-1).

- Do not over-grease the machine (Fig. 4-2). Pumping grease through the joints and on the machine only wastes grease, collects dirt, and quickly makes the whole machine dirty.

- If the fitting will not take grease, move that particular function to relieve any possible stress that might be preventing the grease from flowing through the joint.

4-1

- If the fitting still refuses to take grease, the fitting and/or pin should be removed and cleaned.

- Remember that the pin fittings which are not easily accessible (such as those under the machine) are just as important as the easy-to-reach ones.

- After greasing, clean your hands to remove grease. This is important because it keeps grease off the machine controls.

Hoses and Fittings

The hoses which carry hydraulic fluid throughout the TLB are usually woven from fabric, rubber and steel, and they are made to withstand the tremendous pressures required to actuate the numerous hydraulic functions (2,500 to 3,500 psi). Despite their design, they often become weathered or damaged by hitting or rubbing against objects or other machine parts while working, and the result is usually a blow-out. The operator should learn to spot things such as a slight bulge or exterior deterioration and damage, which indicate a hose is about to fail. Operator awareness of hydraulic system damage due to contact with foreign objects is very important.

Some hose failures are the result of improper installation. If you visualize a hose as nothing more than a flexible pipe, the need for correct installation becomes clearer. Hoses must be allowed to follow their natural flex, and not be unnaturally twisted in one direction or the other. Always use the correct length hose for the job, not too long or too short. Using a hose that is even a quarter-inch short, or the wrong diameter, will guarantee trouble later.

Always take care when installing hoses and fittings. One method to help you install a hose correctly is to mark a white line down the length of the hose and make sure this guide-line is still straight once the hose is in place. If it curves, the hose has been twisted and could be damaged in operation. Some hoses have this guide-line already printed on them.

4-2. DON'T DO THIS.

Another good idea is to keep spare hoses, complete with fittings, ready to go into service should one on the machine fail. This minimizes downtime and guarantees your hoses and fittings will match.

A Word on Tires

The rear tires provide the necessary traction to drive the TLB, and the smaller front tires carry the load when using the loader as well as providing the steering. Each pair is important, and the operator should periodically check to make sure the tires are inflated to the manufacturer's specifications. The rear tires are flotation-type and are inflated to fairly low pressure to provide more traction and to make the loader more effective. Tire pressure may be lowered further when operating in areas where traction is poor. However, the operator should keep the rear tires inflated evenly, to make sure the front loader bucket contacts the ground surface evenly.

Changing Buckets and Bucket Teeth

The bucket is the main component of most of the work done with the backhoe, and depending on the tope of job to be done, different size buckets are available with standard widths ranging from twelve to forty-eight inches.

Changing over from one bucket to another is a fairly straightforward task, but as the saying goes, there are a lot more ways to do it wrong than to do it right.

Bucket Changing Tips

The secret of easy bucket changing is to align the holes of the bucket and the boom perfectly (Fig. 4-3). This will allow the pins to be pushed into place, rather than hammered (Fig. 4-4). Whenever possible, avoid hammering the bucket pins. Hammering distorts the pins and will result in "mushroomed" or generally enlarged ends, which make the process more difficult.

Eliminating the need for hammering is not always possible. New machines, for example, have very tight tolerances, and are generally the most difficult to change. In this case, a large drift pin (preferably made of a softer material than the bucket pin) can be used with a hammer to force the pin out. After several hundred hours of machine use, the pins will wear slightly, becoming easier to install.

There are two "natural" positions whch are critical to the bucket changing process (Fig. 4-5):

4-3

4-4

4-5. The bucket on the bottom shows the natural lying position, and the top bucket shows the natural hanging position.

4-6. Always position the boom straight with the machine when changing buckets. Otherwise the holes will be very difficult to line up properly—compare the incorrect alignment shown in the photo above with the correct alignment in the photo below.

4-7

4-8. This photo shows how to correctly use two hammers when changing a bucket tooth on the shank.

1. Notice the position of the bucket when it is lying on flat ground. When attempting to remove the bucket hinge pin, it is important to place it in the same position as if it was resting on the ground in the "natural position" noted above.

2. Notice the position of the bucket when it is hanging with only the hinge pin attached. This is the correct position to use when attempting to remove the linkage pin (see Fig. 4-5).

It is important to take note of these two positions and use them when removing the pins. The buckets will naturally "find" these positions when the operator installs the pins.

Other useful tips to keep in mind:

- When a hammer must be used, a four to six-pound hammer works best.

- Always have the proper tools on hand necessary for bucket changing.

- Keeping the pins well greased will greatly ease removing and installing them. Keep rags handy to handle the pins, and clean your hands after the change is completed.

- When replacing the boom pin, after the pin has started working its way into the bucket, the boom can be swung slightly while exerting pressure on the pin. This motion will rock the bucket slightly, allowing the pin to align itself with the holes in the boom and the bucket.

- Keep the pins ground to a taper to ease the replacement process.

Correct pin removal/replacement procedure. Always select a smooth, flat, open area to change buckets, and do the actual changing with the boom positioned straight with the machine. This allows easy adjustment of the machine to line up the holes (Figs. 4-6, 7). Then, the following procedure should be used:

1. Position the bucket in the "natural" hang-

ing position. Remove the linkage pin. Retract bucket linkage.

2. Position the bucket on the ground in the "natural" lying position. Make sure the bucket is not hanging from the hinge pin, and also that there is no down pressure from the boom. Remove the hinge pin.

3. Align the boom with the desired bucket. Install the hinge pin.

4. Raise the boom and lower the bucket linkage to the desired position. Install the linkage pin.

Note: This is the only acceptable sequence to follow during the bucket changing process. Any other sequence could cause damage to the linkage.

Changing Bucket Teeth

As the bucket teeth wear, digging resistance will increase. When hard ground is encountered, digging efficiency may suffer. As the teeth continue to wear, they will become thin and will eventually fall off if not replaced. The operator should replace them before this happens; if teeth fall off and digging is continued, the tooth shank will become worn and will eventually be unable to hold a tooth properly. Do not dig with teeth missing from the bucket.

Several tools are needed for the tooth changing process, including an oxygen/acetylene torch; a large, rounded drift pin, or a ball-peen hammer; and a second, preferably larger hammer.

Place the new tooth on the shank in the correct position, and heat one side of the tooth with the torch until it glows cherry-red. Place the drift pin or ball-peen hammer in the correct position, and hammer the now-flexible tooth into the impression on the shank (Fig. 4-8). Repeat the procedure on the other side of the tooth.

PART II.

If you've absorbed the information in Part I, it's time to move on to the more advanced operating principles.

Before we start, understand clearly that it's dangerous to train an operator on a congested construction site. Master the basic control movements off the job, in an open area, using the practice exercises in Chapter 2. Every beginning operator needs the opportunity to learn the most important machine motions without the pressure of completing an assigned task.

Emphasis in the following chapters will be on control — making the machine go exactly where you want and do exactly what you want when it gets there. Good control prevents accidents, increases productivity and simplifies even the most difficult tasks. Without control, no one on the job site is safe.

Many of the tasks covered in Part II require total control — probably more than most operators would believe is possible. Don't be discouraged if it takes you a full year of daily operation to master these movements. Many operators never do learn some of what I'm going to describe here.

As we work through these movements, remember that working conditions vary from job to job and from day to day on the same job. When working conditions are poor or marginal, some of the movements I describe will either be difficult, impossible or downright dangerous. Be safety conscious. Don't try to do anything that seems like more than you can handle. Take it slow and easy. Wait until you have better working conditions before trying the harder tasks.

5. UTILITIES

When digging, one hazard that almost always figures in the planning and execution of the job is the presence of underground utility lines. These include sewer pipes, gas lines, phone lines, electrical cables, and so forth. Being able to locate and avoid breaking utility lines is a very important skill for the operator to develop; otherwise, even the simplest two-hour job can become a costly, drawn-out repair project. Also, in most areas, contractors who are judged responsible for utility breaks are charged for escaped gas or water. Breaking a transcontinental phone cable can become an extremely expensive proposition—up to $1,000 per second for any downtime—although important lines such as these are usually marked with warning signs, and encased in concrete.

The operator will, after a time, develop a "feel" for underground pipes. He will often be able to tell a lot about an underground object simply from the way the bucket makes contact with it. The experienced operator will know if

WORKING RANGE OF THE TLB WHEN LOCATING UTILITIES

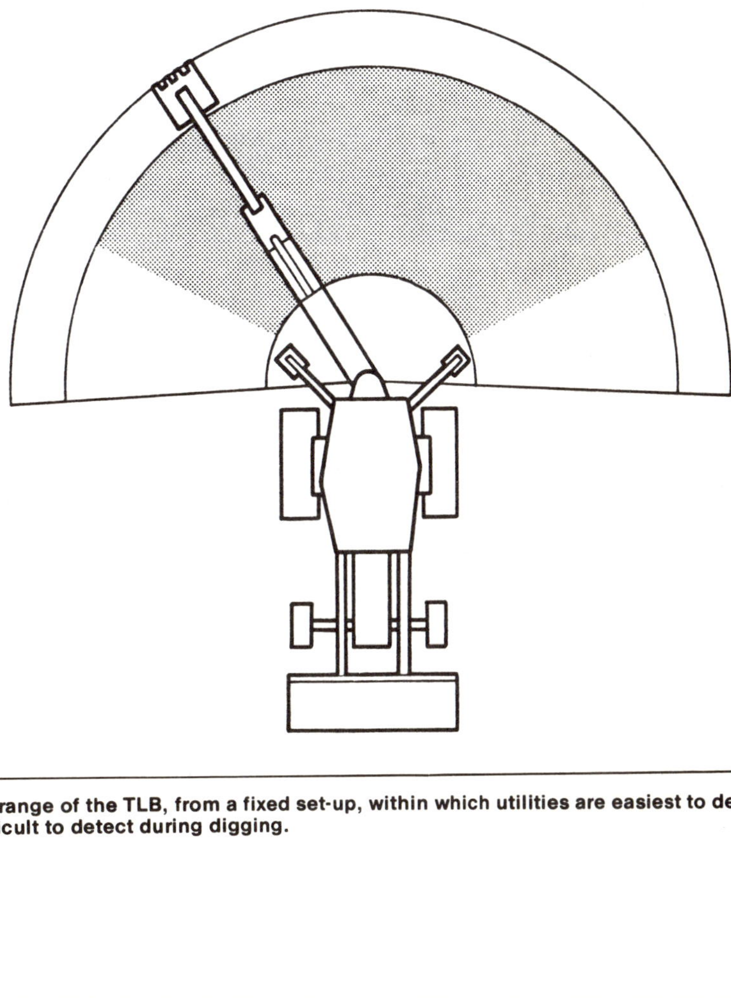

5-1. The shaded area depicts the typical working range of the TLB, from a fixed set-up, within which utilities are easiest to detect. Outside of this range utilities are much more difficult to detect during digging.

he's hitting a rock, a chunk of concrete, or a pipe. If it is a pipe, the really good operator will be able to tell what type of pipe it is, as well as the direction in which it has been laid.

First Step: Finding Them

When looking over the job site, always inquire about possible locations, depths, and types of utilities that may be present. It may often be impossible to get exact location information. Even if they are marked, the markers may not be in the right locations, so there can be many surprises as the job progresses.

The operator should learn to take notice of the "tell tale" signs of underground utilities. Learn to spot such things as the location of gas meters, water boxes, sewer cleanouts, aboveground conduits, electrical vaults and pull boxes, and electrical signs or lights with underground feed. Long, narrow depressions could be the sign of an old trench; in paved areas, trench or pothole patches could indicate underground lines as well.

Also, don't ignore those other signs—the ones that read "BURIED UNDERGROUND CABLE." An old saying reminds us that "Nothing escapes the master's eye," and it is applicable to TLB operation. By being an observant operator, and learning to spot these and other signs, trouble with utilities can be avoided most of the time.

The "Four Stages of Alert"

There are generally four sets of conditions, or stages of alertness, which may exist on any job site with regard to utilities. Knowing how to respond and proceed under any of these stages is vital to handling utilities successfully.

Stage I exists when the operator is working in an area where no utilities are expected to be encountered. This would include open areas, fields, rural sites such as farmlands, or new construction areas. When this is the case, the operator can usually proceed ahead at full digging speed. He will work as quickly as his skills and safety will allow, but he will also be able to stop instantly, if the bucket makes contact with anything unusual. At that point, the operator should use an investigative procedure (described later in the chapter) to determine the nature of the object.

Stage II is where information about utilities is limited, or otherwise incomplete, and the operator simply cannot tell if there are any present. The area will probably have some utilities, but there are no obvious signs of them. This might include most urban locations and many open areas such as parking lots, large lawns, etc. situated in or near any sizable human population. At any rate, Stage II is observed when it is believed there are no utilities present.

Here, the operator can proceed at nearly full speed, but he must be much more careful than in Stage I. Any bucket contact is usually checked out by hand, unless the operator is certain about what he has hit.

Stage III exists when it is likely there are utilities in the area. This would include any time the TLB is operating within the boundaries of "city property," i.e., the street, the sidewalk, or even further in some cases. Signs warning workers about utility locations may be present. In this situation, the operator must exercise great care. He should throttle the machine down and proceed very slowly. This is one more instance when the operator's skills come into play; great care must be used at all times to avoid breaking a line. NOTE: Never dig on public property or within the boundaries of any public easement without obtaining utility location information first.

Stage IV is when the operator knows there are utilities in the area, whether marked or not. In this case, it is important to understand that officials of utility companies often make mistakes about the exact location of their lines; they will usually try to determine a line's location and its depth, but they may be wrong on both counts. If the pipe has been located accurately, however, there is practically no excuse for breaking it. And, of course, there is no excuse for carelessly breaking a pipe that has already been exposed.

Obviously, extreme care must be used in Stage IV situations. The machine should be throttled down to about one-half, proceeding slowly and very carefully. Anything the bucket hits should be investigated by hand.

When the operator is certain there are pipes in the area and the TLB is throttled down, he should proceed with great caution, taking only a few inches with each pass. When the bucket teeth come in contact with something, go back slightly and raise the bucket over the obstruction, and continue digging. At that point a worker should go into the trench to identify the obstruction.

Keys to Pipe Detection

There are three factors which determine whether an operator can successfully locate a pipe through bucket contact and avoid breaking it:

1. Bucket Movement. Watch the bucket carefully for changes in speed caused by resistance from obstructions. When the bucket first contacts a pipe, the bucket speed will change, slowing down as the pipe resists. At that exact instant, the operator must react quickly, stopping the bucket. Relieve pressure on the pipe by backing off in the opposite direction.

2. Control. The operator must be able to move the bucket while maintaining total control at the same time. This can be achieved by following the same practice exercises explained in Chapter 2, "Floating the Bucket" and the "Water Dipping" exercise. Here, however, the object is to go as **slowly as possible.** Performing the Water Dipping exercise very slowly will teach total control of all TLB functions in combination, while practicing the "Floating" exercise will simulate the exact motion used when digging in Stage IV alert situations.

3. Total Concentration. This is the key that combines bucket movement and control into a workable method. Assuming the operator is exercising the greatest control over the machine he is capable of, he must concentrate both on the bucket and the ground in the immediate area of the bucket. By getting the "big picture" of the excavation area, the operator can usually pick up signs of ground movement, or a change in soil content indicating an old trench. Stay alert to any possible sign of previous excavations—if they are discovered while the trench is shallow, extra care can be taken as the trench is deepened.

Variables That Determine Pipe Breakage

There are a number of variables that help determine whether or not a pipe is broken in the course of TLB operations. These can include:

Soil Conditions. If soil is hard or rocky, "feeling" the pipe will be difficult. Detecting pipes in soft or sandy ground is much easier.

Depth of Pipe. If a pipe is fairly close to the surface (two feet or less), hitting it will sometimes cause earth movement in the immediate area, or on the sides of the trench. When this happens, stop and investigate. This warning sign becomes less useful at greater depths, since ground movement is less likely to occur, unless the soil is very soft or loose.

Type of Pipe. Steel pipes (three inches and larger) and concrete storm drains are the only types of pipe which are fairly easy to locate with little risk of breakage. Cast iron, copper, clay, transite and large-diameter plastic pipes are also **possible** to locate and identify by feeling them with the bucket, depending on the operator's skill and other variables discussed here.

Diameter of Pipe. Small-diameter pipes (three inches or less), and wires are very difficult to feel except when ground conditions are very soft and movement is noted, as discussed earlier. Generally, though, these types of lines will have to be located by hand.

The angle of the pipe when encountered. Ideally, the best way to locate a pipe with the least chance of breakage is to set up the TLB so the pipe crosses the trench exactly perpendicular to the machine. Finding the pipe to be situated perpendicularly will equalize the bucket force at all points touching the pipe, and the chance of breaking it will be reduced. Of course, it usually doesn't happen that way; the pipe will often lay at some angle relative to the trench. Even finding the pipe to be exactly parallel to the trench (i.e., running directly under the machine) is not as bad as hitting the pipe at an angle. Depending on the severity of the angle, contacting the pipe this way concentrates the backhoe's power at only one point on the pipe. This greatly increases the risk of breakage, and should be avoided whenever possible.

Age and condition of the pipe. It is often hard to determine how long some pipes have been buried, and the amount of deterioration that has taken place over time. Most types of concrete, clay and plastic pipes generally do not deteriorate as rapidly as other types. A rusted metal water or gas line, on the other hand, can start leaking simply by disturbing the soil around it. To avoid breakage or leaking in these cases, carefully locate the lines and dig around them by hand.

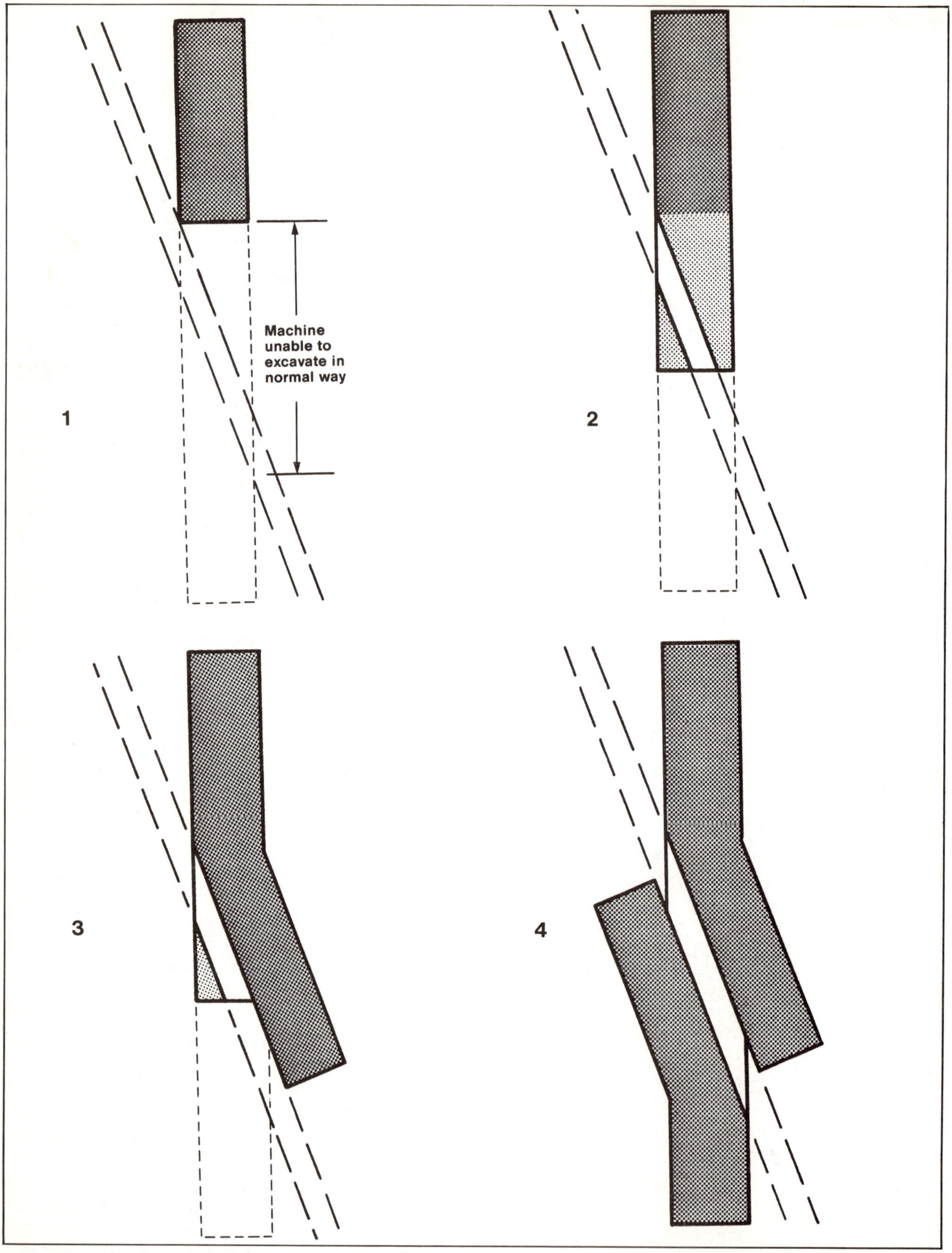

5-2. In this series of diagrams the shaded area shows how to dig around a pipe once it is encountered. When pipes are found to be at an angle relative to the trench, the operator must carefully expose the pipe first, and then proceed to dig around it.

Investigating Possible Utility Lines

This is the procedure commonly used when the operator encounters an unknown bucket obstruction. First, he must remove the bucket from the trench without breaking whatever it might be, and then determine what it is. If it is a pipe, go on to the following procedure. If the object is a rock, a root or debris from an old excavation, digging may continue as before.

Digging Around Pipes. Once a pipe has been located and identified, the operator must monitor its exact location until the excavation is completed to the point where it is out of the machine's working range. The best way of monitoring its position, obviously, is to be able to **see it at all times.** If the pipe is covered with spoil, it should be re-exposed by hand as quickly as possible under the circumstances. Often, a laborer will place his shovel in front of the pipe to mark its location. Although a common shortcut, this is a dangerous idea which should be discouraged.

5-3. Often hand labor will need to be used, once the bucket has encountered a pipe, to carefully expose it.

5-4. Example of how the bucket carefully works around a pipe once it is exposed, to avoid breakage.

Instead, the pipe should be cleaned by hand so the TLB operator can clearly see it. (Hand labor often is required in this procedure.)

If the pipe is plastic, copper, clay tile, or if the line in question is electrical wiring or telephone cables, hand labor should also be used to prevent major repair problems.

Digging Around Utilities

Learning to operate around utilities is an important element in the operator's repertoire of control skills. Even experienced operators occasionally break a pipe because, as noted earlier, it is often impossible to locate them with any precision. But learning to distinguish a pipe from a rock or other buried object is another acquired talent that comes only from experience on the job site. The sooner it is developed, the less chance there will be of delays and repairs caused by utility accidents.

Once a pipe is successfully located, the TLB operator's job is to dig around it. Expose as

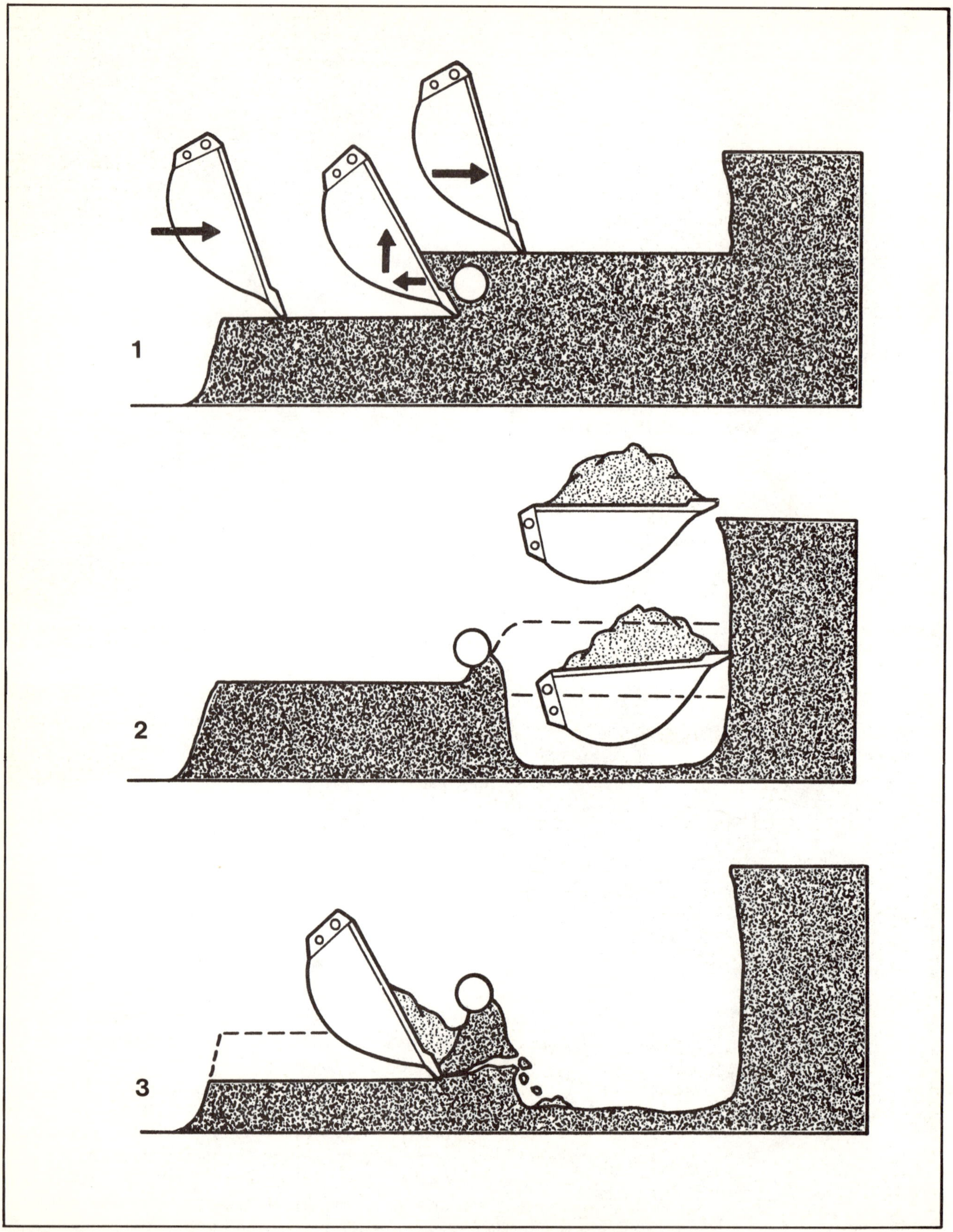

5-5. DIGGING AROUND UTILITIES: 1) During the digging process, when contact is made with an unknown object, pressure is relieved and the bucket is backed off, and then raised over the object. The obstruction is checked out by hand, and if it is a pipe, it will be exposed so that the operator can see it from his operating position. 2) Here, material on the near side of the pipe is carefully removed. 3) Then, by carefully pushing material under the pipe, it is then freed of the surrounding material.

5-6 (above) and 5-7 (below) show a water pipe near the surface that was inadvertently broken during a concrete breaking and removal job. When such a break occurs quick action is necessary, such as shutting the water off at the main, before repairs can begin (Fig. 5-8, next page).

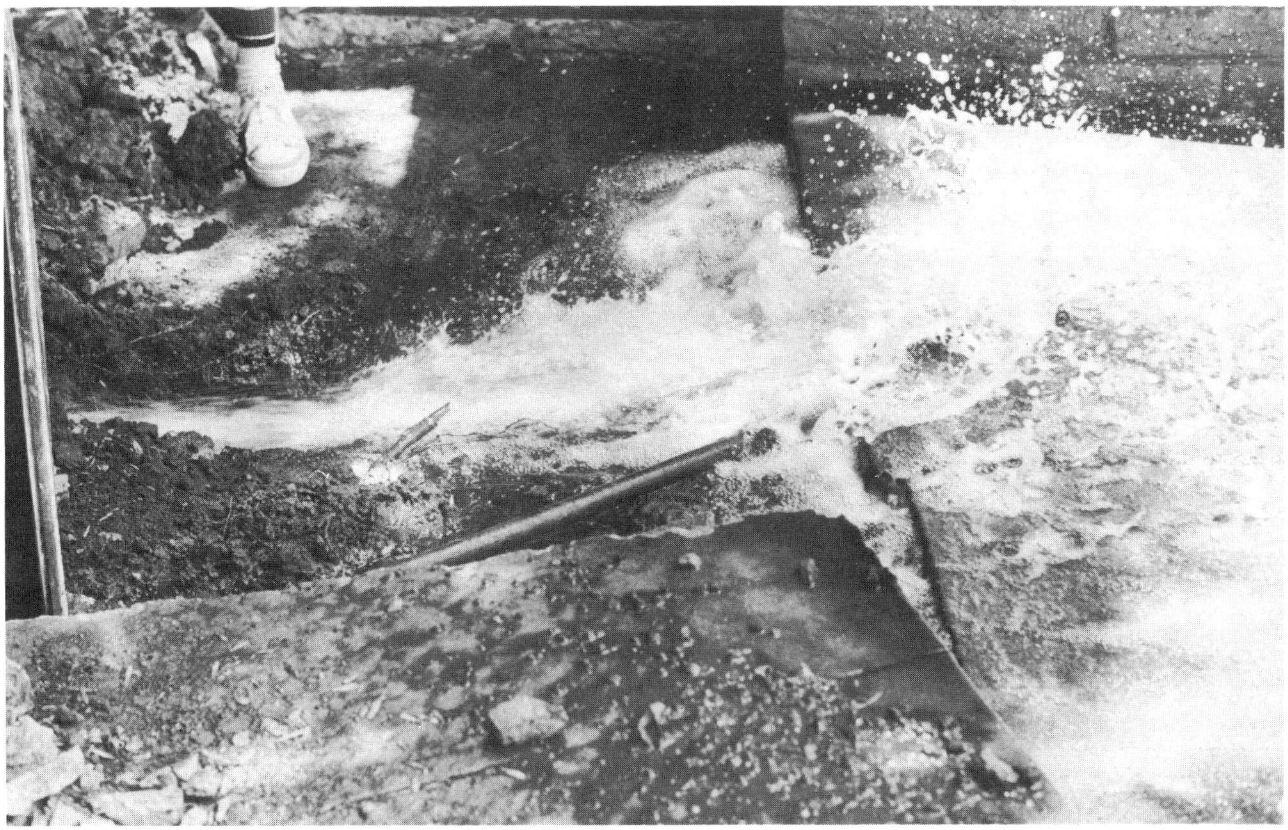

much of the pipe as possible, using the machine as well as hand labor, until it is **clearly visible from the operator's position.** Once past the pipe, proceed excavating the trench as usual, taking great care not to strike the pipe with the bucket or any part of the machine.

When a Pipe is Broken

If a pipe happens to be broken (as it sooner or later will) during the course of TLB operation, the operator will know it fairly quickly. If it is a water line, the water will make its presence known, and the operator will be able to smell, see and hear the gas escaping from a ruptured gas line. If any kind of electrical or telephone cable is severed, the exposed wires will be visible, jutting from the sides or floor of the trench.

When a pipe or other line is broken, stay calm and **think** about what to do. Action should depend on what type of line has been broken. Notify the workers and/or utility company officials whose job it is to make necessary repairs. If it is a gas line, **do not shut down the TLB before getting it clear of the area.** Attempting to start up the TLB in a gas cloud could ignite the fumes. If a water main has been broken, move the TLB away from the flood plain. If service to a residence or residences is involved, there is usually a shutoff valve at the street. It would be advisable to locate these shutoff valves prior to digging, so quick action can be taken if a break occurs. These may include water control boxes, gas shutoff valves, etc.

In all cases, notify the affected utility company immediately of what has happened, so repairs can be made as soon as possible, and work can be resumed.

5-8

6. PLANNING THE JOB

When the TLB operator first arrives on the job site, he needs to make a mental inventory of the tasks expected of him. He should walk the job site, developing a job scenario in which he visualizes a) the types of tasks he will have to perform, and b) the order in which he will proceed, to complete the entire job as efficiently as possible.

The operator should consult the person in charge of the job site to determine the scope of the work involved. Will it involve excavation for footings, utility lines, sewers, basements or septic tanks? Will material have to be stockpiled, transferred, removed or loaded onto trucks? Will land have to be cleared or graded? Will roads need to be built? Will the job require demolition?

These are the major tasks the TLB operator will find himself confronted with most often, and coming up with a logical, efficient job plan will make it easier to move from task to task.

Important points to consider when consulting the person in charge include utility locations and possible problems arising from them; placement of spoil; and bucket requirements (several bucket changes may need to be planned in advance). Discuss anything that may be relevant to the work expected of the TLB, and do not start to dig until each sequence of the excavation has been planned, to avoid digging the machine into unmaneuverable situations and causing delays.

In addition to determining the digging sequence, the operator should also be mindful of coordinating his work with the other trades on the job site. He will also have to plan locations for stockpiling, and how to transfer the material to the stockpile area when necessary. As the TLB's work nears completion, the operator should think about the following phase of construction, and plan the wrap-up of his work accordingly.

On some jobs, where shuttle operations will require large quantities of material to be moved about, it may be necessary to build or grade access roads, or build ramps over curbs, footings, trenches, etc. (See chapters on ***Grading and Compaction*** and ***Loader Technique***.)

Placement of Spoil

The main consideration in placement of spoil on the job is, of course, what will eventually be done with the material. If it is to be used for back-filling, it should be placed in an area that will facilitate back-filling when the time comes. Or, if it is to be trucked out, put the spoil in an area that is accessible to the loader and the truck itself, to facilitate loading. (See chapter on ***Set-up*** for methods aiding in spoil placement.)

Layout

Be wary when the person in charge says, "Well, start out over here in this and dig me a ditch over to that there." Trenching is rarely such an inexact process. Yet, when people try to explain what should be done, they are often unclear. Misunderstanding someone's instructions can create problems throughout the entire job. The best way of avoiding this kind of misunderstanding is to have the area clearly marked. Occasionally, those in charge of a job may avoid clearly marking the excavation. This might be due to simple laziness, or more likely, they may not know exactly where they want the operator to dig. But it's not up to the TLB operator to decide where to put a trench. That information can only come from the person in charge, and he must communicate it to the operator as exactly as possible.

Precision Job Layout

In some types of excavations (particularly footings), a precise layout is the most important element in completing the job with professional results. Because of the high cost of concrete and labor, it is very important to get the footings cut to the exact specifications, and in the exact locations, the first time.

Whenever the operator is called on to perform work of this exactness, he must have a thorough understanding of the job and its dimensions. Discuss the job with those in charge until all aspects of the excavation are clear, and each step is understood completely. Go over the plans and make sure of the layout. **Remember:** when you start out right, you end up right; start out wrong, and you'll end up wrong.

To be structurally sound, walls must be built directly on the footing to the approved plan. Since the wall will occupy a fixed, exact location, the footing must also be in the exact location. When marking the area to be dug using a ball of string or other straight edge, along with lime or some other white powder, care must be taken to keep the marks clean, straight and narrow. Applying the marking material from a few inches above the ground will assure a precise layout.

Marking both sides of the footing is essential if the footing to be dug is wider than the bucket being used. It's also a good rule to mark both sides whenever precise excavation is required.

Sometimes, excavations for plumbing or electrical lines will also require precise layouts. In plumbing, these layouts will usually be called for when planning ground-floor lavatories, floor spills, or any place where pipes will come up through the concrete in an exact spot. The same requirements apply for placement of electrical conduits.

Job-Planning Tips

As a review of the material included in this chapter, as well as a suggested procedure to follow when inspecting the job site prior to beginning work, the TLB operator should keep the fol-

6-1. Precise layout lines on the sides of the excavation are an integral aspect of correctly planning any precision digging job.

lowing general tips in mind when planning the job:

- Always select the shortest and safest route through the job site.
- Check terrain and soil conditions that may reduce traction or require changes in the direction of travel.
- Be aware of the following obstacles, and check for them: trees, buildings, pavement, equipment, terrain embankments, personnel, and utilities (either underground or overhead).
- Consider the nature of output desired and expected of the TLB, the amount of earth to be moved, and the time available to perform the work.
- Place temporary reference markers to indicate obstacles to be avoided: grade stakes, utility pipes or wires, thin concrete, objects that could cause tire damage, etc.
- Discuss the nature of the work with other personnel involved in the job as necessary.

In conclusion, no operator likes to do a job a second time because of poor planning. As we've seen, poor planning can be very costly when precise excavation is required, in addition to causing needless delays. When planning a job, the good operator is the one who has all his moves planned out well in advance, so that as one part of the work is completed, he will know what the next step is, and be able to move right into it without error and delay.

7. SET-UP

Introduction

Set-up of the TLB is a basic fundamental, one which applies to every job the machine will perform. Correctly positioning, or setting up the machine prior to digging, must be understood to operate the backhoe efficiently and professionally.

Proper set-up consists, basically, of lining up the machine with the trench at the optimum digging distance. The main elements of a correct set-up are: 1) the center line of the hoe, which must be positioned correctly to cut the trench exactly to the layout lines; 2) positioning of the entire machine in such a way that easy stockpiling of spoil in the desired area is possible; 3) making sure the machine is level, in order to produce a plumb trench.

When making plans for stockpiling, the operator must consider areas that will permit easy access to the machine or other equipment such as trucks, in order to make back-filling or spoil removal easier. Positioning of the machine in relation to the stockpile is done by placing the "arc of the swing" in the correct position. This is done by angling the machine itself in one direction or another, relative to the excavation (Fig. 7-1).

The Tripod Set-Up

The backhoe can only be operated efficiently when it is positioned on a solid base. The weight of the machine should not be supported by the tires. Instead, the machine is best supported by a **tripod set-up,** consisting of the **front loader bucket** and the **two stabilizers** (Fig. 7-2). There are numerous advantages to using this method, while problems created by it are few.

Advantages. For whatever reason, many operators choose not to lower the front loader bucket during operation. Quite simply, efficiency is lost in this manner. When the weight of the tractor and loader is supported by the front tires, the machine will bounce slightly. The operator will, in turn, bounce slightly also, and this motion is transferred directly into the controls (usually causing the machine to shake even more). The backhoe cannot be operated to its maximum potential in this manner. However, this can be easily corrected by using the tripod set-up.

The loss of control produced by the extra motion also costs the operator a degree of safety. This alone is good reason to use the tripod method.

Another point to consider is the tendency of the backhoe to be constantly pulled backward during operation due to the tremendous forces generated by the hydraulic system. The tripod set-up, however, puts more weight on the stabilizers. This has two positive effects: 1) With the front bucket off the ground, the hydraulic digging force tends to lift the stabilizers off the ground, freeing the machine to roll on the front tires. The tripod set-up puts more weight on the stabilizers, keeping them firmly on the ground. 2) The extra weight also increases traction, which further prevents backhoe slippage. **In this position, the operator can summon the greatest machine resistance against being pulled backward due to the digging forces.** This

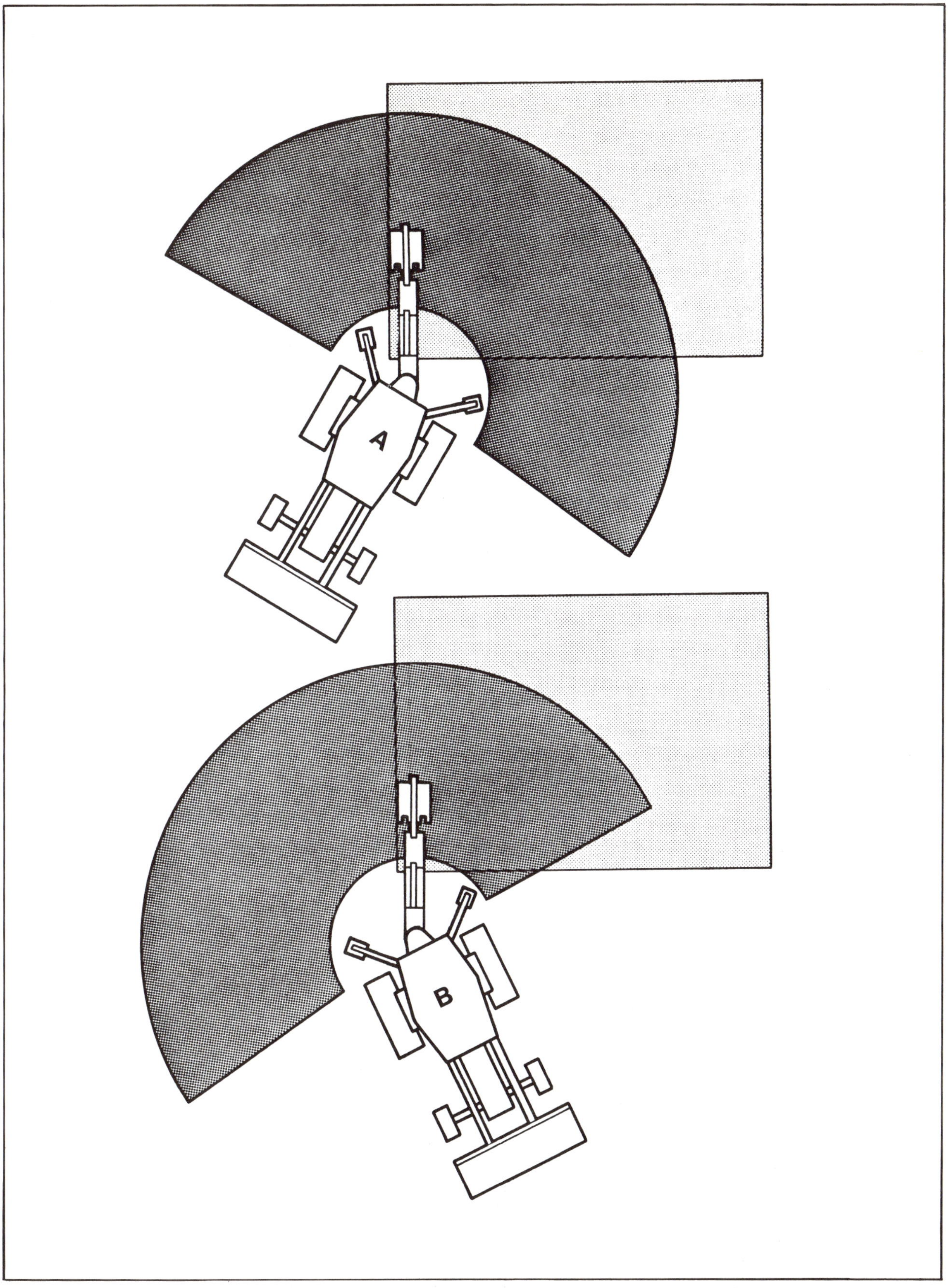

7-1. SET-UP 'A': In this example, the desired depth of the excavation is ten feet deep. Keep in mind that the spoil must be kept back away from the excavation. In this position the operator will rapidly run out of space for spoil placement. SET-UP 'B': In order to increase the spoil placement area the arc of the swing is adjusted by re-positioning the machine at a different angle to the excavation.

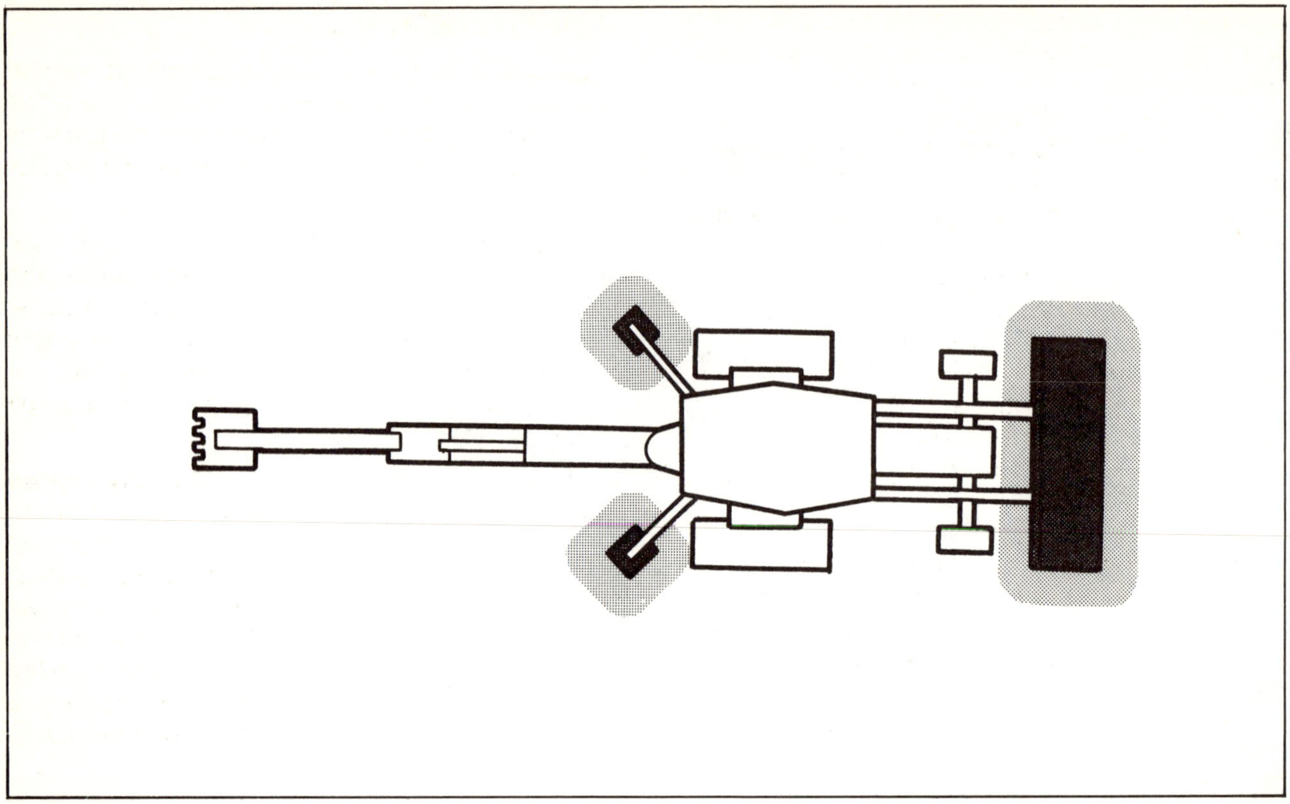

7-2. Illustration shows the basic use of the tripod set-up utilizing the loader bucket and two stabilizers to provide a working platform for the backhoe.

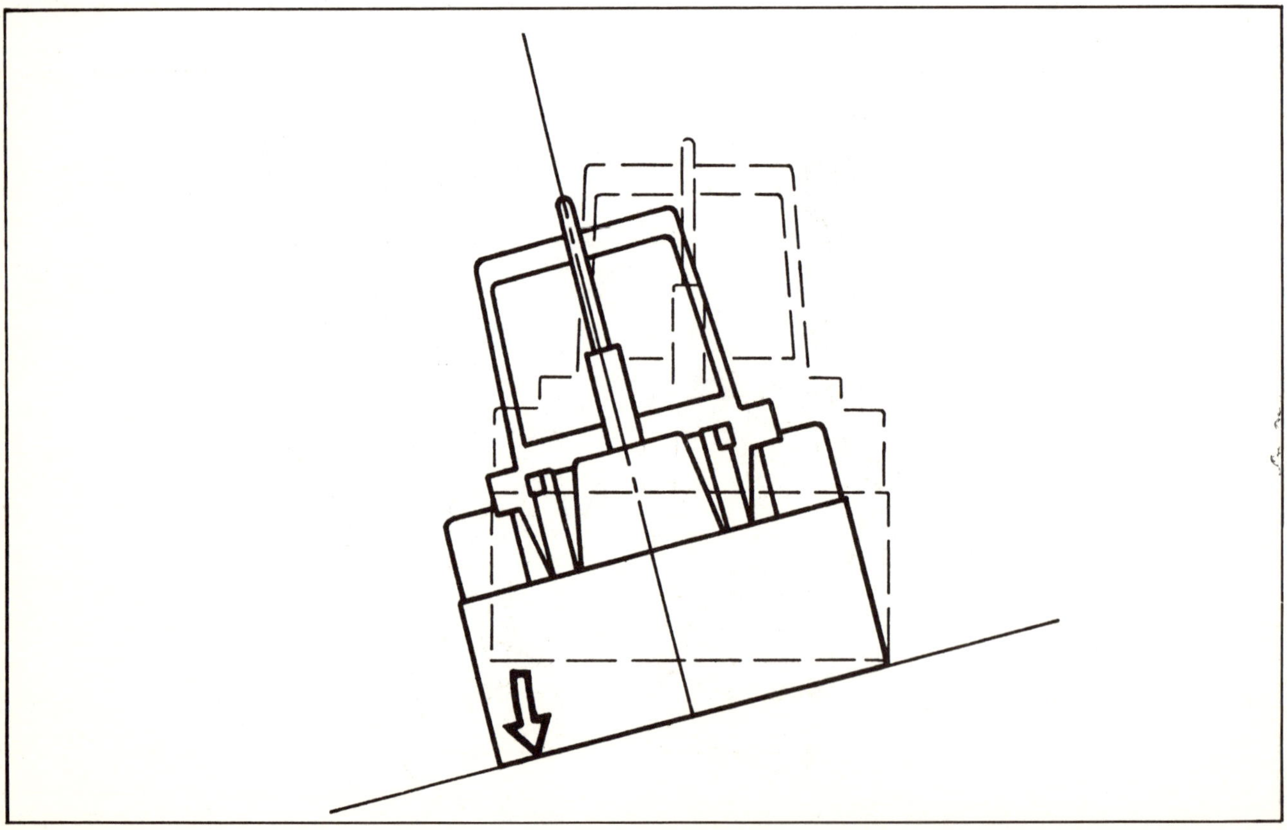

7-3. When the bucket is lowered on sloped ground, the machine will be forced in the direction perpendicular to the slope.

concept should be understood and practiced, and used to the operator's advantage whenever possible.

One other advantage of the tripod set-up is that with the loader bucket firmly lowered on flat ground, the TLB is stabilized. Any time the loader bucket is off the ground, stability is decreased.

Problems. The tripod set-up is not without some drawbacks. When operating on concrete or a delicate surface such as tile, masonry or stonework, direct front bucket contact is undesirable. In these situations, it may be necessary to place wooden planks or plywood between the contact areas, or have the front bucket raised while working.

When operating on hillsides or anywhere the area under the front bucket is not level, the machine will be forced in the direction perpendicular to the surface (Fig. 7-3). To avoid this, lower the bucket only until it makes enough contact to stabilize the machine.

Using the tripod set-up may sometimes cause problems when there is insufficient area to rest the front bucket. These situations present themselves rather infrequently, however, and can be remedied by raising the loader bucket up until it no longer presents a space problem.

The Pivot Method

When using the tripod set-up, the TLB may be pivoted 25 degrees in either direction, using the lowered front loader bucket as the pivot point as the machine maneuvers to one side or the other (Fig. 7-5).

Repeating the pivot procedure over and over would result in the machine making a circle, with the loader bucket/pivot remaining in the same place. While the need for this maneuver might not arise very frequently, understanding the pivot concept is important because it is used often in different situations.

This pivot method is used most often when aligning the machine to the excavation, when connecting two trenches, or when crossing trenches. In fact, crossing trenches without using the front loader bucket as a stabilizer/pivot is not recommended, as it would be unsafe. Whenever the front loader is firmly down on flat ground, the machine is stabilized and tipping or rollover is controlled while the machine moves sideways for a new set-up.

There are many job situations where the pivot method is useful. Here we will examine one common situation, and some of the possible set-up

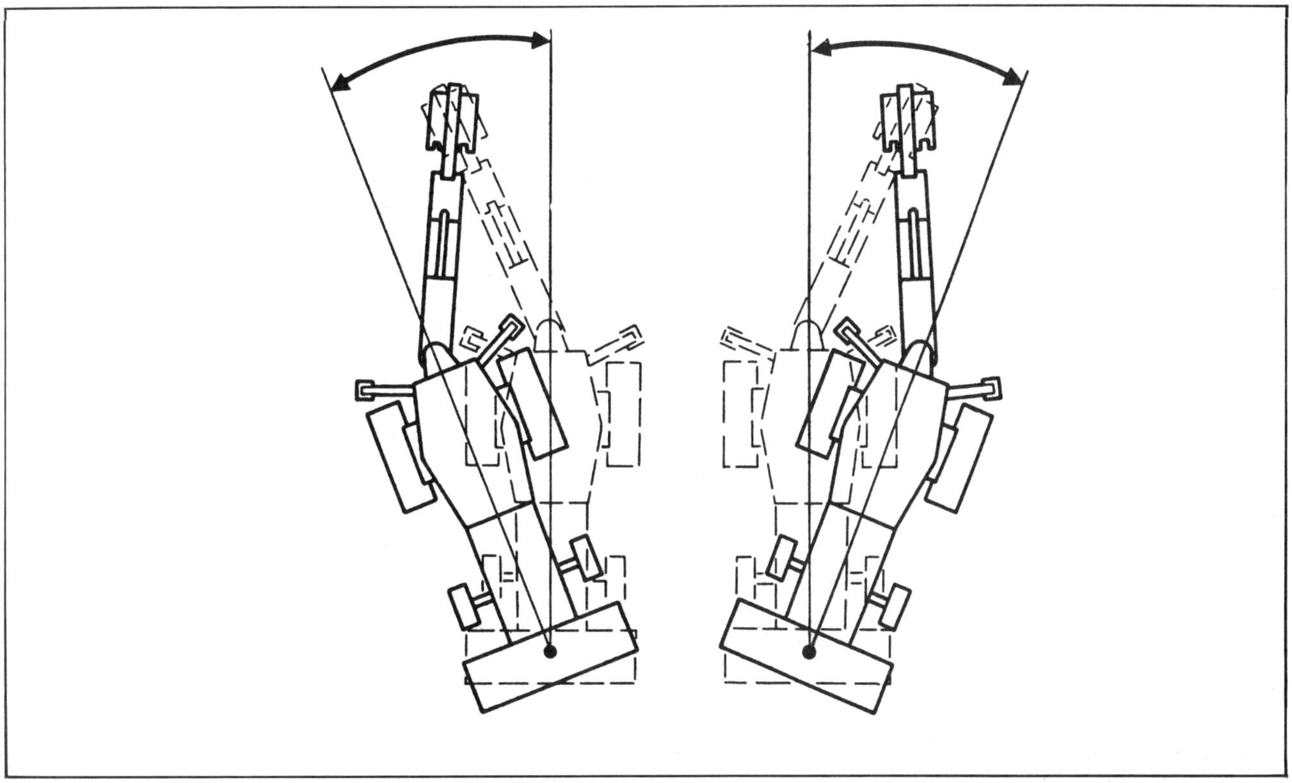

7-5

7-7

7-8

procedures. Often, the operator will be asked to dig a trench between two buildings. To accomplish this task, the trench must be started at the ends first, and dug until the two trenches meet at a **planned** mid-point and a single trench is produced. If the trench is shallow enough, and the area stable enough to support the weight of the TLB, there are three possible set-ups to choose from.

Set-up #1. As seen in Fig. 7-6, the machine is driven to the desired position and the stabilizers lowered. Fig. 7-7 shows the bucket lowered to a stable area at an angle that will actually increase the stability of the machine as the pivot is performed. After the pivot is completed (Fig. 7-8), the stabilizers are lowered onto stable areas and the swing pin is positioned directly over the center of the trench.

Set-up #2. Using the loader bucket as a stabilizer is useful in many ways (Fig. 7-9). The TLB was driven into the position shown here, with the loader bucket supporting its weight. Again, the stabilizers were lowered and the bucket positioned at an angle advantageous for the pivot. Fig. 7-10 shows the position directly after the

7-9

7-10

pivot has been made. By comparing these two pictures, the movement of the machine becomes clear. Of course, such maneuvers are only attempted on very stable ground.

Set-up #3. The loader bucket is again used to support the weight of the machine as it is slipped to the opposite side of the trench. The stabilizers are lowered, and the operator takes his position at the backhoe controls (Fig. 7-11). Because this particular pivot is too much to make in one "jump," the machine is moved to the point where the stabilizer is nearest the trench and still on stable ground (Fig. 7-12). The bucket is then re-positioned and the second "jump" is made (Fig. 7-13). Again, notice the position of the stabilizers and the swing pin. In this particular case, the operator needed to end up on the other side of the trench, so this method worked very well (Fig. 7-14). It should also be noted that the "arc of the swing" is positioned differently in each of these set-ups (Figs. 7-8, 10 and 13).

Precision Set-up

Precision set-up is called for most often when the backhoe must dig footings (see ***Footings***). It simply involves aligning the machine exactly with the excavation. As seen in Fig. 7-15, the backhoe is extended to the farthest point de-

7-11

7-12

7-13

7-14

7-15

7-16

sired, and the operator adjusts the swing so the bucket touches the layout line at the desired point. (Once the swing is in the correct position, it is not moved until an alignment check is completed.) Then the bucket is lifted up and pulled in to the closest part of the layout (Fig. 7-16). If the bucket contacts the layout at the desired point, the backhoe is perfectly aligned to excavate that particular section of the job.

If the backhoe does not line up correctly, it can be pivoted slightly to the side, with the procedure repeated until the machine aligns perfectly with the excavation. With practice, this technique becomes a habit and is performed quickly. After the machine is set up properly, the operator must be **constantly aware of the machine's alignment.** The **bucket** can be knocked out of alignment by some type of obstruction such as a rock, pipe, a piece of concrete or even a root. In addition, the **machine** can be pulled out of alignment by the hydraulic digging forces, as we have noted. This usually occurs in hard or rocky soil conditions.

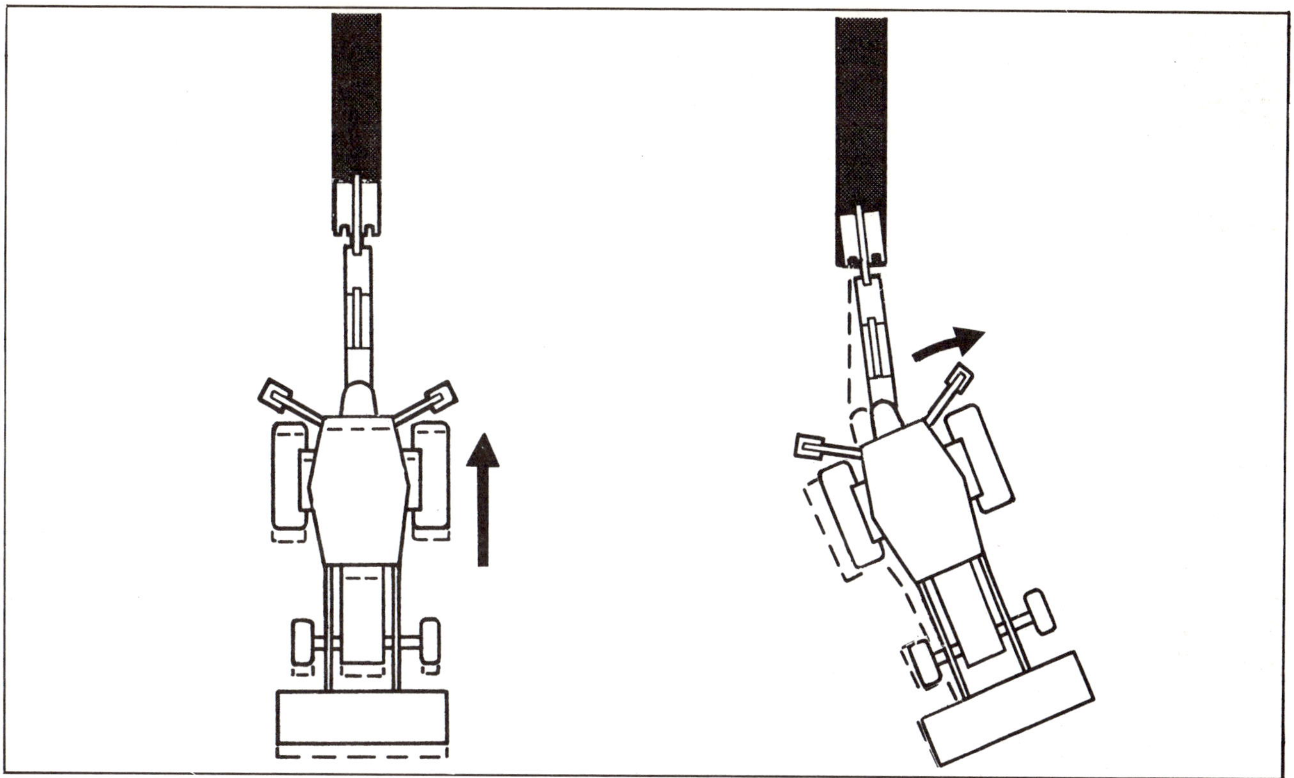

7-17

The best way to avoid the machine being pulled out of alignment is to position the machine straight with the excavation. This means that if the machine is pulled backward by the digging forces, it will still be aligned with the trench (Fig. 7-17). Any other set-up will cause the backhoe to be pulled out of alignment by the digging forces.

In deep or large excavations, the operator must pay particular attention to the sides of the trench, leveling the machine to keep it plumb, as plumb sides are important to precision digging.

Set-Up Techniques

In order to examine different set-up procedures, we will depict three different examples of how to set up the TLB to dig a common pier footing. The procedure that the operator will choose on the job will depend on the working area available and the desired spoil placement area. The excavation in all examples is the same —twelve feet square and four feet deep.

Fig. 7-18 illustrates the first approach to accurately digging this kind of excavation. Because the sides are the most important element of pier excavating, the set-up must aid in the precise cutting of the sides that will follow. To save time and make the job easier, two sides of the excavation will be dug from a single set-up, set-up #1. The machine is pivoted to the side slightly, and is perfectly aligned with the other side. When following this procedure for pier excavation, the sides are always dug first, and the center area is cleaned afterward. After the bulk of the dirt is removed, the bottom of the footing is fine-graded and cleaned carefully, without over-digging. Spoil is then pulled into one corner of the excavation and removed. When completed, the TLB is moved to the other side of the excavation, set-up #2. The same procedure—digging the sides first, and **then** digging the center out—is repeated.

The second approach to setting up for this type of excavation is illustrated in Fig. 7-19. The operator moves the machine to the first set-up position and digs the first two sides. The TLB is then moved to set-up #2, where the other two sides are excavated. At that point, the perimeter of the footing is completed, and the center is then dug to the desired depth or elevation, and cleaned.

The third example is illustrated in Fig. 7-20. This procedure requires three separate set-ups, but works well when very precise results are desired. The operator starts in the first position,

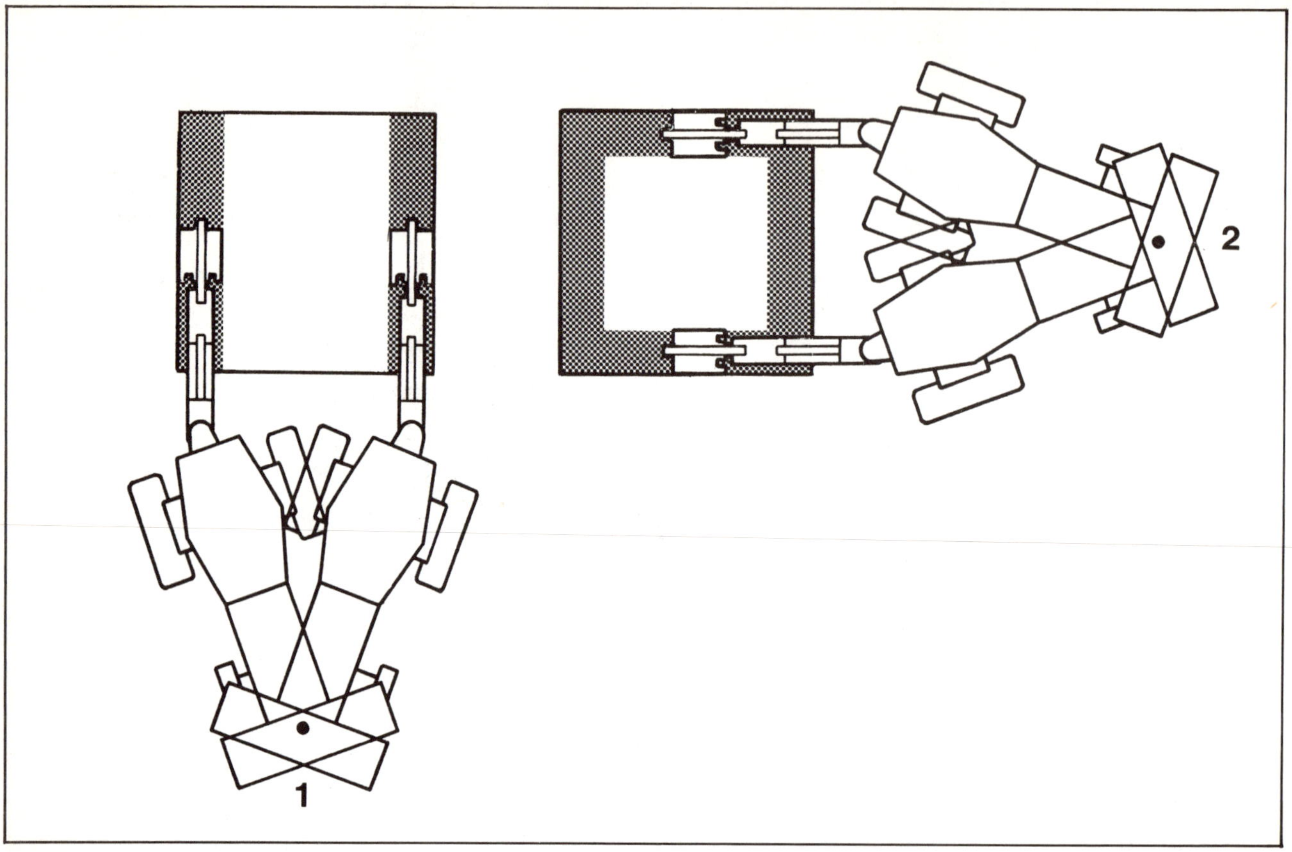

7-18

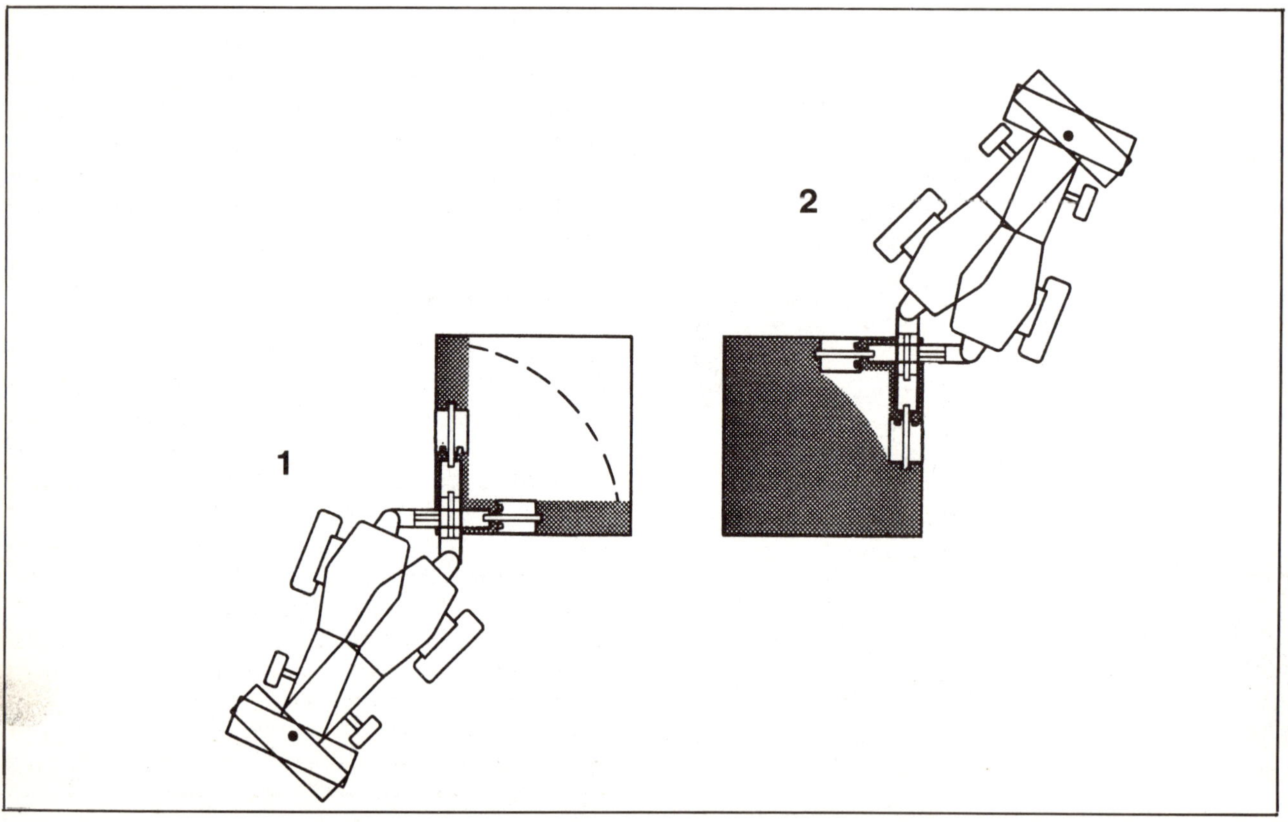

7-19

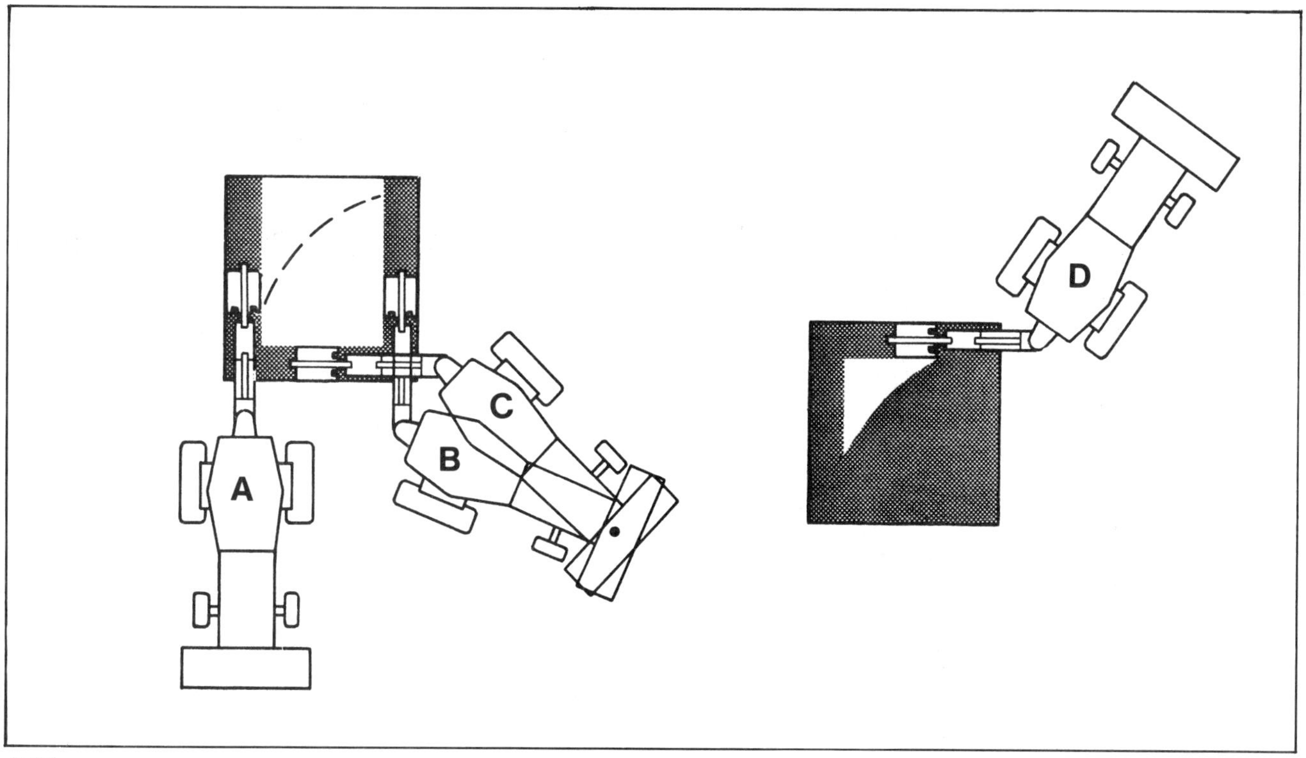

7-20

7-21

from which side "A" of the footing is dug to the desired depth. The machine is then moved to the second set-up position, from which sides "B" and "C" are dug. At this point, three sides of the excavation are dug to the correct depth.

As the operator begins to clean the center, the three completed sides give him good visual aid in cutting and cleaning the center. When work is completed, the machine is then moved to the third set-up position. The operator then cuts the final side "D" and cleans the remaining spoil from the excavation.

Conclusion

In this chapter, we've tried to explain the two major components of backhoe set-up, namely the tripod set-up and the pivot method. The two procedures are interrelated and both enhance the operator's control of the backhoe. Whether the job is a precise excavation, as for a footing, or just a simple trench, proper set-up is essential to getting the work done correctly and efficiently. Like other aspects of TLB operation, set-ups, once they are practiced, become second nature to the operator. When care is used in pivoting the machine, the operator will discover many ways the backhoe can be used to perform a number of jobs, in many circumstances, that he might not have thought were possible—such as the unusual set-up seen in Fig. 7-21.

8. FOOTINGS

The excavation of footings is often considered to be the most difficult operator skill to fully develop, because of the degree of precision required. In the process of excavating footings, the backhoe operator actually builds a concrete form as the dirt is removed. Footings must be dug in the exact location, according to the dimensions in the job plan. If they are dug too large, concrete will be wasted; if they are dug too small, the reinforcing steel will not fit. If this occurs, the contractor faces increased costs or risks failing an inspection by building authorities. In this chapter, we'll examine job layout and precision digging techniques that will enable the operator to excavate footings correctly **the first time,** since there is no second chance once digging has started.

Footing Types

Trench-type footings. These are generally the easiest types of footing to dig, as they usually have only two requirements: 1) proper set-up of the machine, to ensure correct location of the

8-1. Here the operator cleans the side of the trench, which will prevent materials from falling back into the trench, as well as ease the work of the laborer.

trench; and 2) achieving the correct depth of the trench, including a smooth, flat bottom.

Before starting to excavate, consider all aspects of the job ahead of you. Plan the layout as thoroughly as possible, to get a good mental picture of the excavation sequence.

Important points to consider include the amount of spoil that will be created by digging, and where it will be placed. The operator must also plan a step-by-step order in which to proceed with digging, so that all the footings may be dug easily and an exit route provided for the TLB.

When excavating trench footings, it is best to dig each section of the trench as long as possible, going down quickly to within about one foot of the desired elevation. Then, the operator should use long, level passes with the bucket, taking out about two inches at a time, checking elevation at both ends when necessary, until the desired grade is reached. Another good idea, when the operator gets to within about six inches of the desired grade, is to clean the sides of the trench with the bottom of the bucket, creating a clean trench area and a safer working environment (Fig. 8-1).

Sometimes it may be necessary to start the footing at both ends, working toward a meeting point. Plan ahead where the two trenches will meet, so there will be enough room to use the pivot method if necessary (see ***Set-up***).

8-2. The first step to accurate excavation of pier-type footings is to dig out the "sides" first.

Pier-type footings. These are the footings usually called for when large, multi-story structures are being constructed. They are basically large concrete platforms, either square or rectangular in shape, that will eventually support the weight of the entire structure.

Requirements for pier-type footings include straight, smooth sides that are plumb, and in exactly the right location, as well as a smooth, level bottom at the correct elevation. The best approach to digging pier footings is to dig out the sides first, making sure they are straight and correct, and then simply removing the earth remaining in the "center" of the footing (Figs 8-2, 8-3, 8-4, 8-5).

For example, let's consider a typical pier footing, to be dug twelve feet square and four feet deep. Excavating a footing of this size will require removing 21.3 cubic yards of material. Since the standard twenty-four-inch backhoe bucket has a capacity of about one-quarter cubic yard, it will require about eighty-five full buckets to remove the required amount.

8-3. Once one side is complete, move to the opposite side and repeat the procedure.

8-4. After all four sides are dug out the operator then removes remaining material in the "center" of the footing making sure that the bottom reaches the desired elevation.

8-5. Photo shows the results of a correctly-dug footing.

8-6

8-7

8-6, 8-7, 8-8: Top photo shows the area of the desired layout; middle photo shows the area cleaned of debris and smoothed for the layout; above, the finished layout.

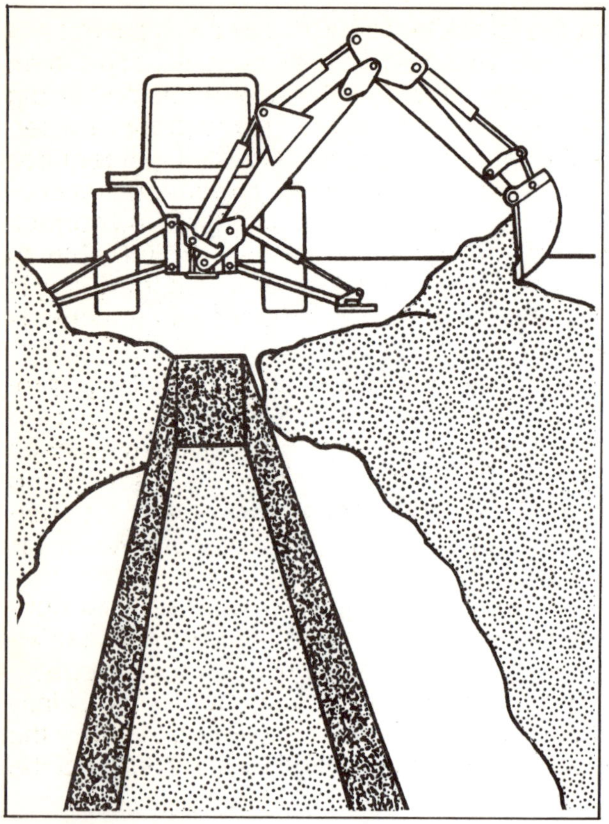

8-9. A correctly dug pier footing will have straight sides and a smooth, flat bottom . . .

Of course, just removing spoil from the area is not enough to finish with a professionally excavated pier footing. Before removing the dirt, think about the requirements mentioned earlier: 1) straight smooth sides that are plumb and in exactly the right location, and 2) a smooth level bottom at the correct elevation.

Layout

As we have already noted, an accurate layout is essential to digging to the correct dimensions and in the right locations. Since dimensions and locations are so important to the excavation of footings, it follows that an accurate layout is indispensable when footings are to be dug.

Layouts for footings must be clearly marked with narrow, precise markings of lime, or any material that can be clearly seen (Fig. 8-8). Layout lines should be extended past the limits of the excavation, so the location is still visible after digging has begun. When digging trench footings, be sure to mark **both** sides of the trench, especially when the footing width is wider than the bucket.

It should also be noted that a clean area for a layout is a must for jobs involving any amount of precision. Cleaning the layout area ahead of

8-10 . . . and the top of the excavation will be cleaned of spoil.

time will ease the operator's task by simplifying precise layouts and making maneuvering of the TLB easier. (See Figs. 8-6, 8-7 and 8-8.)

Precision Digging

Footings require the most precise digging the TLB operator is capable of. And yet, this type of digging is difficult only when the operator does not understand, or is not aware of, the basic requirements of precision digging. They are: 1) digging to the exact perimeter in the right location, and 2) digging to the desired elevation.

To meet these requirements, the operator must have an **exact** layout; some means of checking elevations at any time during excavation; and a laborer to assist the operator during excavation. These factors, combined with a knowledgeable operator, usually result in clean, precise footings (Figs. 8-9, 8-10).

Elevation Importance and Checking Methods

The depth, or elevation of the footing is one of the two requirements for expert results. The main idea is to dig to the depth required, but not any deeper. Some jobs are more exacting than others, and it is sometimes very difficult to dig exactly to a certain elevation and produce a perfectly flat, level bottom. It is a skill that requires total operator concentration, great bucket control and a deliberate style of cutting and correcting as the work progresses. Being able to periodically check the elevation while digging is also important, and there are several methods that are widely used.

The most common grade-checking procedure involves a builder's level, which is capable of pinpointing a level plane at any point on the job site. Once the instrument is set up and leveled, the grade is then checked, and the operator knows how far he has to dig to reach the bottom of the excavation (Fig. 8-11).

A "story pole" is often used with the level, but a simple tape measure will suffice when the reference plane is within reach of the person checking the grade. When the job is especially exacting, it is absolutely necessary to check the grade as digging proceeds, at both ends of the ditch.

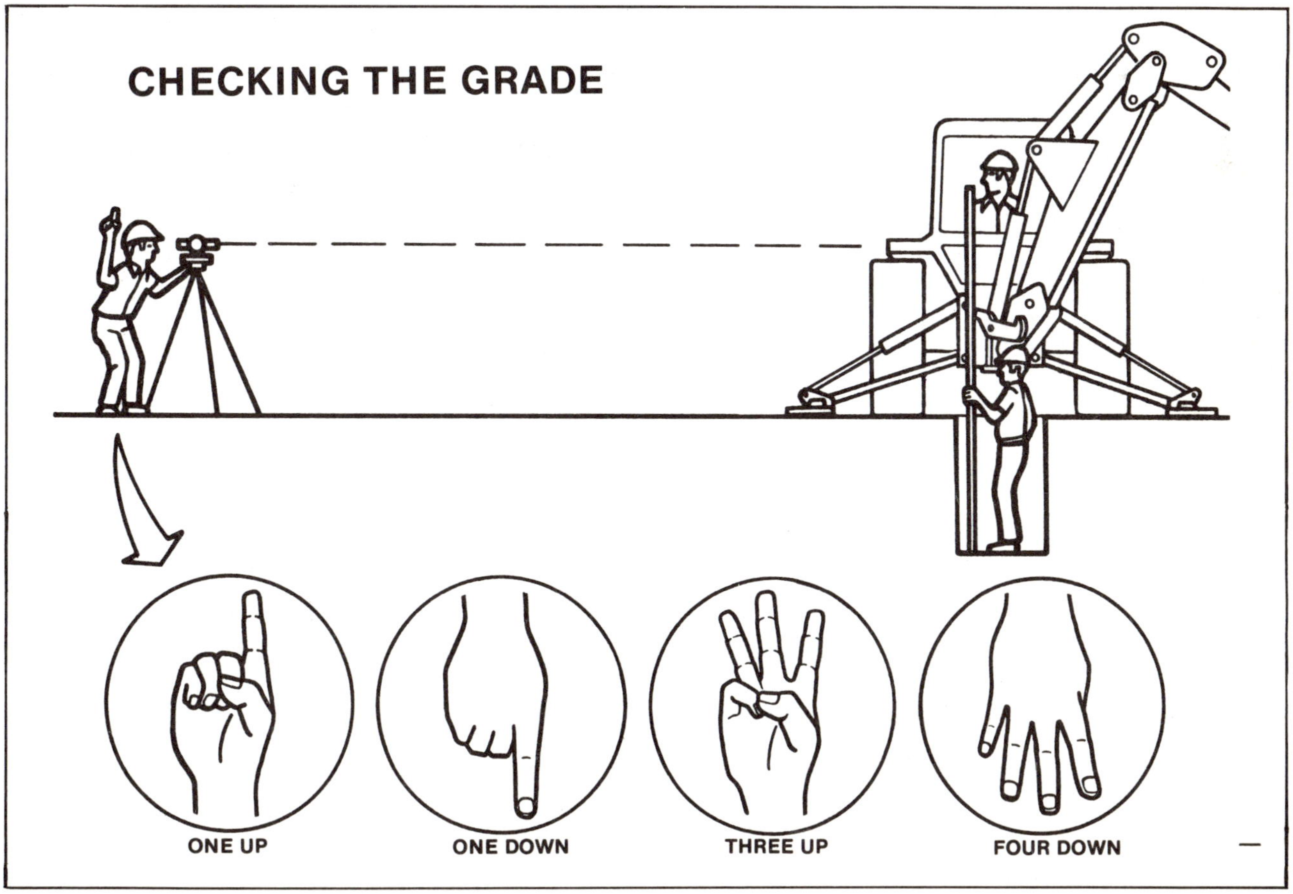

8-11

Measuring the **depth** from either side of the trench is usually unacceptable, because it is not an accurate grade check. However, this method is all right for other types of excavations, such as plumbing or utility lines, and reduces the need for constant references to elevation.

Laborer Involvement

When digging precise footings, it is advisable to use a laborer to assist the TLB operator in checking elevations and in other tasks. One common practice is to have the laborer clean the excavation of loose spoil, to provide a more solid base for the concrete to be poured on. The sides of the footing are also cleaned to prevent dirt from falling in later, making it safer for the men who will be working in the excavation after the backhoe's work is done (Fig. 8-12).

When the operator and the laborer work together as a team, digging efficiency is increased. The laborer aids the operator by cleaning the grade and providing a visual grade-checking function. As he pushes loose dirt toward the machine, cleaning the bottom of the excavation, the laborer indicates to the operator how much more to cut, and warns the operator against over-digging. The operator then removes the spoil, making the laborer's job easier.

Having a team of an experienced TLB operator and laborer makes digging footings more efficient, and greatly reduces the chance for error and delay. This teamwork greatly enhances the likelihood that the footings will be dug correctly the first time, which is of prime importance.

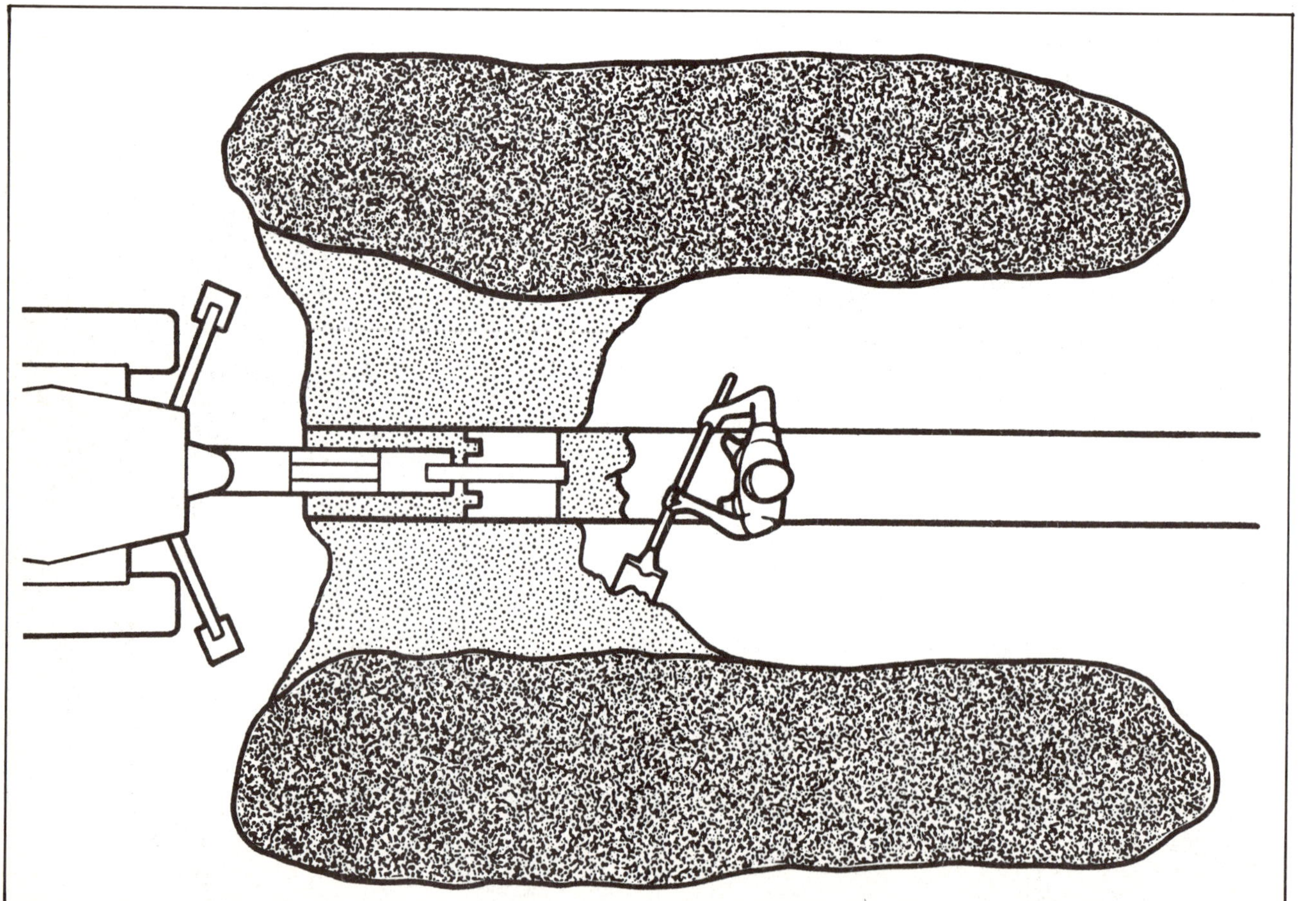

8-12. The laborer faces the machine while it is working, cleaning the spoil from the sides and pushing the spoil forward in the trench for the machine to remove.

NOTE: This sequence is intended to show the order in which to proceed when excavating footings, and not the correct procedure itself. In an actual job situation, a laborer would help clean the footing and assist the operator in checking the grade as digging progresses. If no laborer is available, the operator should then proceed in the same manner depicted here, digging as best he can and cleaning the work area of spoil.

8-13. At this point the operator has dug two sides of the footing and has piled the spoil outside of the excavation area.

8-14. As the operator moves to the next set-up the machine is angled to allow spoil placement outside the excavation area.

8-15. As the trenches are dug they are graded as cleanly as possible on the bottom.

8-16. The inside as well as the outside of the work area is cleaned of spoil to provide a usable working area.

8-17. After cleaning the digging is resumed.

8-18. The operator sets up the machine for digging the final side.

8-19. Beginning the surface cut.

8-20. The footing is completed, and the spoil removed.

9. JOB SITE CLEANLINESS AND CLEAN-UP

Introduction

As we have discussed in other areas of this book, job site cleanliness can actually enhance the TLB operator's performance. We should also note that a smooth, well-graded working area is the key to production when using the front loader. Maintaining a clean job site also allows much easier maneuvering of the machine at any time.

The clean-up methods described here are generally used toward the end of the job, or to periodically clean the work area. Clean-up is required at the end of most jobs. It may involve back-filling the trench, moving or stockpiling materials, or loading spoil onto trucks. In the case of paved areas, clean-up may require removal of excess material so the area is clean enough for sweeping. How to best complete these tasks usually depends on the type of surface the TLB is working on.

Shuttle Operations

When an area is to be utilized for any repeated shuttling back and forth of material or machine, it should be graded smooth before starting to transport material. After several trips are completed, the area will be compacted, and should be back-dragged to fill any depressions or smooth out rough areas. As work progresses, the operator should continue to back-drag the rough areas, keeping the shuttle area smooth and polished.

Some operators may overlook this technique and dismiss it as a waste of time. On the contrary, it is time well-spent—efficiency is increased in many ways: a smooth working area allows the machine to travel quickly around the job site without losing control; it prevents loose material from spilling out of the bucket; and operator fatigue is lessened—he doesn't have to physically and mentally fight the machine to control it, and his alertness is maintained throughout the day.

Even when shuttle operations aren't involved, a clean working area is important to performing nearly any TLB job. In some situations, the benefits may not be readily apparent; in the long run, however, there are far more cases when a clean area will simplify the work greatly.

On nearly all job sites for commercial and high-rise construction projects, spoil is stock-

9-1. Dragging spoil back and away from the trench.

piled outside the immediate building area in a designated stockpiling area. In the case of plumbing or electrical excavations which require back-filling, material is shuttled or trucked in from the stockpiling area later, when it is required.

As spoil is being removed from, and returned to the excavation area, the main consideration is to maintain a clean, smooth shuttle road from the work area to the stockpiling area.

Spoil Removal

To remove spoil from the trench area, first drag it back away from the trench (Fig. 9-1). This prevents any spoil from falling back in, and at the same time it cleans the area next to the trench. If there is enough room, it should be pulled back about ten feet. This will provide room for side-spill from the bucket without material falling back into the trench during clean-up.

If the TLB operator makes a habit of smoothing his working area, soon the entire job site will be cleaner and more productive for all trades on the job. If the operator's loader skills and finish-grading skills are sharp, he can make the job site look polished to the eye (Fig. 9-2). In the long run, the ability to produce these kinds of results will make one's services as a TLB operator more valuable.

Importance of Excavation Cleanliness

The TLB operator and his function are not independent of the other workers and roles on the job site. As the backhoe creates trenches and other excavations, these areas will become the working location for other construction trades such as electricians, plumbers, laborers, ironworkers and concrete finishers. As the backhoe digs excavations for any of these purposes, the methods and procedures followed by the operator will determine the productivity and safety of the work area for all trades on the job that will follow. The TLB operator must keep in mind that the fact he is digging a trench is often matched in importance by how he has prepared the work area for the other trades that will also work on the job.

9-2

9.3. As the loader moves forward from this position the bucket will cut the high places as desired.

9-4. Back-dragging will fill the low areas.

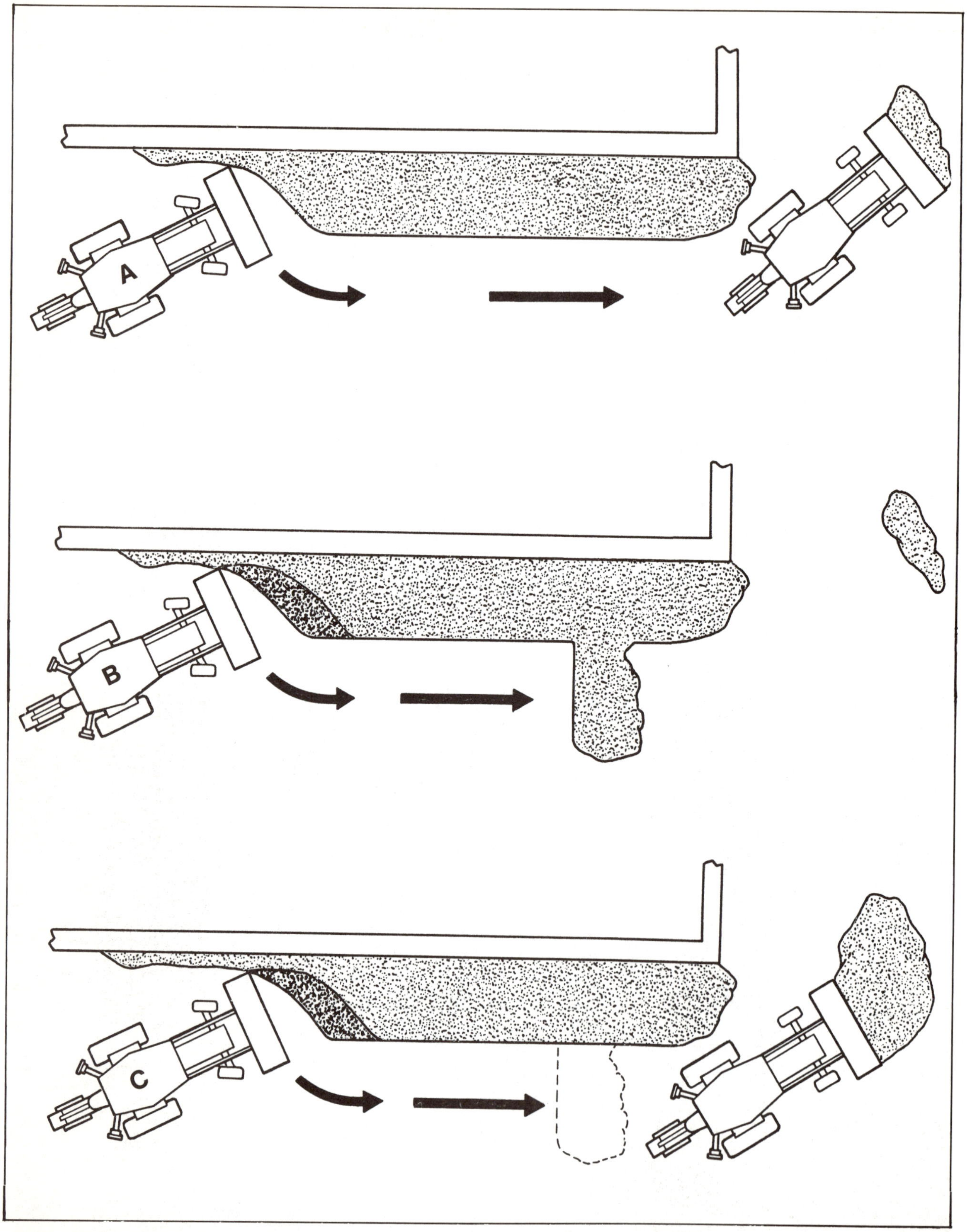

9-5A. On the first pass the bucket is filled and the edge of the pile is cleaned. The material is then stockpiled in the desired area. 9-5B. On the remaining passes the machine is initially positioned at a slight angle. This will prevent the tires from running over the edge of the pile causing the machine to tilt. Instead of pushing one bucket at a time, the full length of the clean-up section, every other bucket can be pushed part-way. 9-5C. Then the material is pushed the rest of the way on the next pass.

Clean-Up

Sod or lawn surfaces. When working on sod, the main goal is to do as little damage to the surface as possible. If the topsoil is wet, it may be too soft to support the machine's weight without some damage. In such conditions, a half-full loader bucket could sink the front wheels, making movement impossible and creating quite a mess. If conditions are ever this unstable, it may be necessary to use the backhoe for back-filling. In this situation, it is advisable to distribute the machine's weight through the stabilizers and the tires. This provides a larger area of weight distribution, and less chance of the machine creating ruts or holes in the lawn.

When cleaning up in sod areas, the loader bucket should be **flat**, with slight down pressure applied. With the bucket flat, it will not dig into the sod; and with a little practice, it will be possible to clean off large, flat areas right down to the grass, without digging in. Areas that are not flat are somewhat more difficult, and should be finished by hand after the bulk of the spoil is removed.

Another aspect to keep in mind is to apply power to the rear wheels carefully, to avoid spinning them. It may even be necessary to take

9-6. The material next to the edge may be left for the final clean-up.

9-7. By positioning the machine in this manner the edge can be cleaned while keeping most of the machine weight away from the excavation.

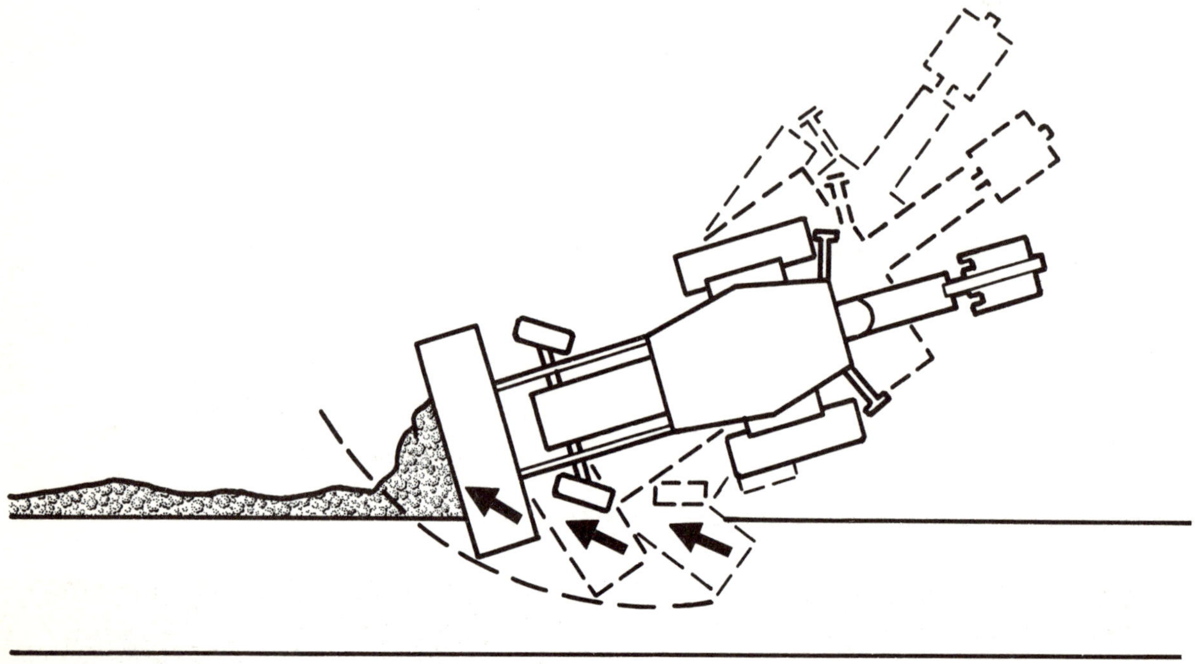

9-8. By making several sweeping passes the edge can quickly be cleaned without pushing material back into the trench.

smaller "bites" of the spoil pile to prevent spinning the wheels unnecessarily.

Dirt Surfaces. Clean-up of dirt surfaces is less complicated than cleaning sod areas, and usually consists of cleaning off the surface to the original grade, or simply leveling and grading the area. Like sod clean-up, the bucket should be flat on the ground during actual cleaning. If a cut is desired, the bucket should be tipped slightly down, and more weight applied by lowering the loader more. To prevent over-cutting, the bucket should be re-set in the flat position.

The simplest and best method of cleaning up in dirt areas is to cut down the high areas on the bucket's forward passes (Fig. 9-3), and filling the low areas while back-dragging (Fig. 9-4). This method wastes no motion, and the job site should quickly take on a "polished" look.

Paved Surfaces. The object of cleaning paved surfaces is to scrape the area clean and either consolidate the material or load it out on trucks. This task is most often called for when trenches have been dug in a street, parking lot or similar paved area. In these situations, the operator should take care to **not** fill the trench back in, and to limit the spread of the dirt on the paved surface. To accomplish this, the clean-up procedure depicted in Fig. 9-5 works quite well.

As the clean-up procedure nears completion, the remaining material may be next to the excavation (Fig. 9-6). By positioning the TLB as seen in Fig. 9-7, and making a series of sweeping passes, the edge of the excavation is cleaned safely without pushing the spoil back into the excavation. This motion is continued until clean-up is completed (see Fig. 9-8).

9-9. As the pile becomes smaller it will no longer serve as an effective back-up.

9-10. By pulling the corners back . . .

9-11. . . . the center of the remaining pile will serve as a back-up.

9-12. This procedure is repeated several times until the pile is cleaned up.

Pile clean-up. Another common clean-up task is the removal of a stockpile or spoil from an area. When a back-up is available (some object to push the load against to facilitate loading), the job is much easier. In many cases, however, there is no available back-up, so this pile clean-up method should be followed.

When filling the loader bucket from a large pile of dirt, the bucket will fill easily and quickly. As the pile becomes smaller, the back-up (the pile itself) becomes smaller, and filling the bucket becomes more difficult. When the pile is reduced to only two or three cubic yards of material, the most the operator can expect to load into the bucket is less than half of the remaining amount. In order to load even this much, the pile must be positioned by the operator in such a way that he can take advantage of it as a back-up. Figs. 9-9 through 9-12 illustrate how the operator can "groom" the pile to provide a constant backup to facilitate loading.

Note: It is very important that the operator avoids pushing the pile to the extent that he loses his back-up. Once the back-up begins to move, the amount of material he is able to load into the bucket will not increase.

As loading continues, the operator will take smaller and smaller amounts, until only about one-third of a yard of material remains. At that point, the remaining material can be loaded by hand, and the area cleaned by hand if desired.

10. GRADING AND COMPACTION

Introduction

In this chapter we will discuss both rough-grading and finish-grading techniques as well as the principles and methods of soil compaction. Although the TLB is not designed specifically for grading, it can be used efficiently as a grading machine in some situations. Grading, simply, consists of **cutting** down the high places and **filling** in the low places.

Required Skills

To be effective at grading, the TLB operator must develop and practice two abilities. First, he must possess expert loader skills. This includes the ability to cut, grade and maneuver the machine in such a way that certain cuts and other moves can be "attacked" from optimum angles. Second, the operator must be familiar with the grading process, and how to complete a job. To get the most out of the TLB's capabilities when performing these tasks, the operator must pay attention to these skills and develop them through practice whenever possible.

Grading Techniques

As we mentioned, grading is simply a matter of cutting the high places and filling the low ones to produce the desired grade. By visualizing the entire grading job **as it will look** when it is finished, and at the same time looking at the job **as it is,** the cut-and-fill areas will become clear to the operator. As he travels the job site during the grading process, "bucketing" material from the cut areas to the fill areas, he should continually study the cut and fill sections. He should also make constant mental comparisons between the desired finished grade and the actual grade. As this evaluating process is repeated from several different vantage points on the job site, the required movement of material will become very clear. This process is repeated as the grading job progresses, until the required grade is achieved.

By following this procedure, **cutting** can usually be done to the exact elevation the first time, requiring only a light pass at the end to level the area. **Filling,** on the other hand, often requires compaction of the earth, and then refilling and leveling to the desired grade.

Loader Grading Techniques

Rough grading with the loader is usually possible only in areas that are relatively flat, because of the TLB's greatly reduced stability and maneuverability on slopes. Attempting to use the loader for any type of slope work is not advisable, since the results will usually be unsatisfactory, and the procedure is generally unsafe.

It should also be noted that the grading process may include a general clean-up of the job site. The operator may be required to separate debris like rocks, roots, broken concrete or asphalt, waste construction materials, etc. The materials, once separated, are then usually stockpiled for loading onto trucks, or dumped into trash bins on the site.

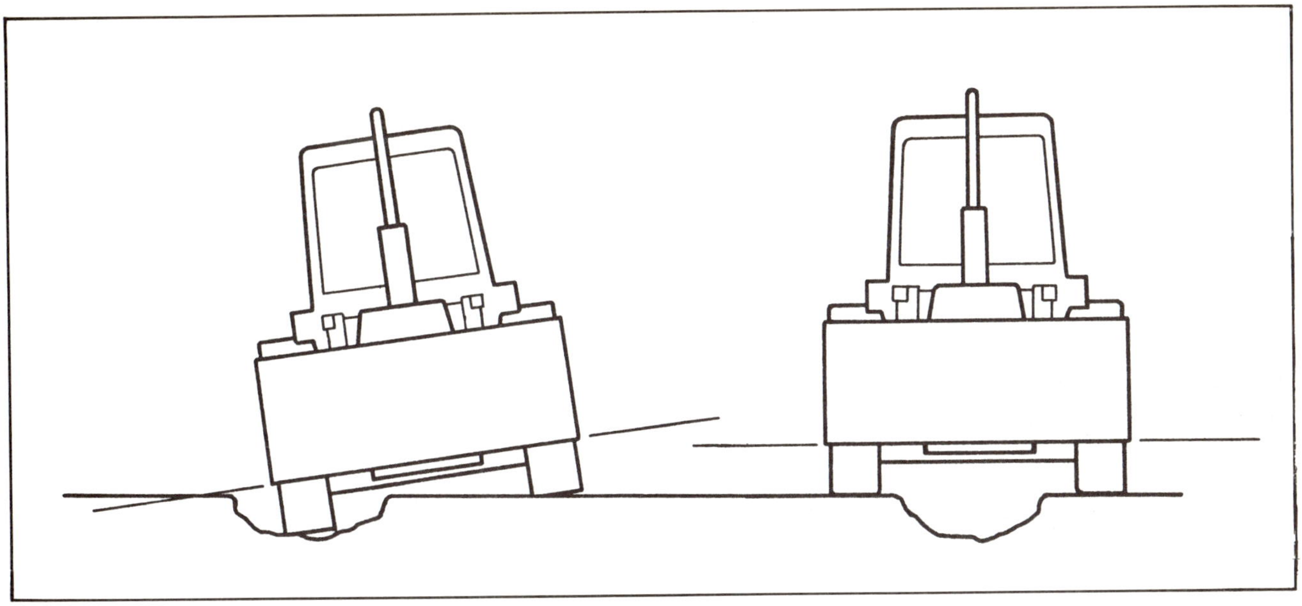

10-1

The **range of jobs** the operator may be required to perform can be broken into three categories: 1) cutting an area and either loading the material onto trucks or stockpiling it for later loading; 2) cutting from one area and filling another area elsewhere on the site; 3) filling an area with material from a stockpile, or with material that has been trucked in. The common element running through all these types of tasks is that grading materials on the job site will often need to be adjusted (stockpiled, moved about, etc.) to achieve the desired grade.

In contrast to the frequent teamwork called for by the TLB operator with other workers, grading is a job where the operator is usually "on his own." After receiving and discussing job instructions with the person in charge, it is then a matter of the TLB operator alone "at the controls." It is his job to plan and execute the necessary movement of material to achieve the desired grade. It is also his job to come up with a well thought-out plan of exactly how the work will proceed, step by step. **Controlling** the TLB to produce exactly the desired results is the operator's responsibility alone.

During both the planning and execution phases of this type of job, the operator should do his best to avoid wasted motions. He should move directly from one phase to the next: cutting, transporting material, compacting, and then filling. He should also correct the grade as the job progresses, and keep the cut and fill areas, as well as any access roads, polished and free of obstructions. Accomplished operators will blend these tasks together into a procedure that will always have a goal, or purpose, or at least ease completion in some way.

One example of this efficiency effort is to keep the loader bucket full whenever possible to improve the TLB's compaction capabilities. This can be done by driving over the fill areas while bucketing material from the cut areas.

As work proceeds, with one area being cut and then another being filled in, the visual results won't become apparent until the bulk of the material has been transported into or out of the area, whichever the case may be. Only when the polishing or finish-grading has begun, and corrections made, will the desired results appear. At that point, what may have appeared as a disorganized mess will finally begin to make sense.

Cutting

Starting the cut. The first step for the operator is to envision the desired cut in his mind. Take notice of the desired grade and the degree of slope, if any, and compare it mentally to the existing grade and/or slope. The key to the cutting process is that the attitude or angle of the bucket cutting edge is determined by the angle of the area where the rear wheels are making contact (Fig. 10-1). Because the cutting edge angle cannot be adjusted, the operator must instead make necessary adjustments in the work area, so that the bucket will make the cut at the desired angle or level.

Once the machine is at the desired angle or level of the cut, the cutting process can then easily continue at that grade. There are several

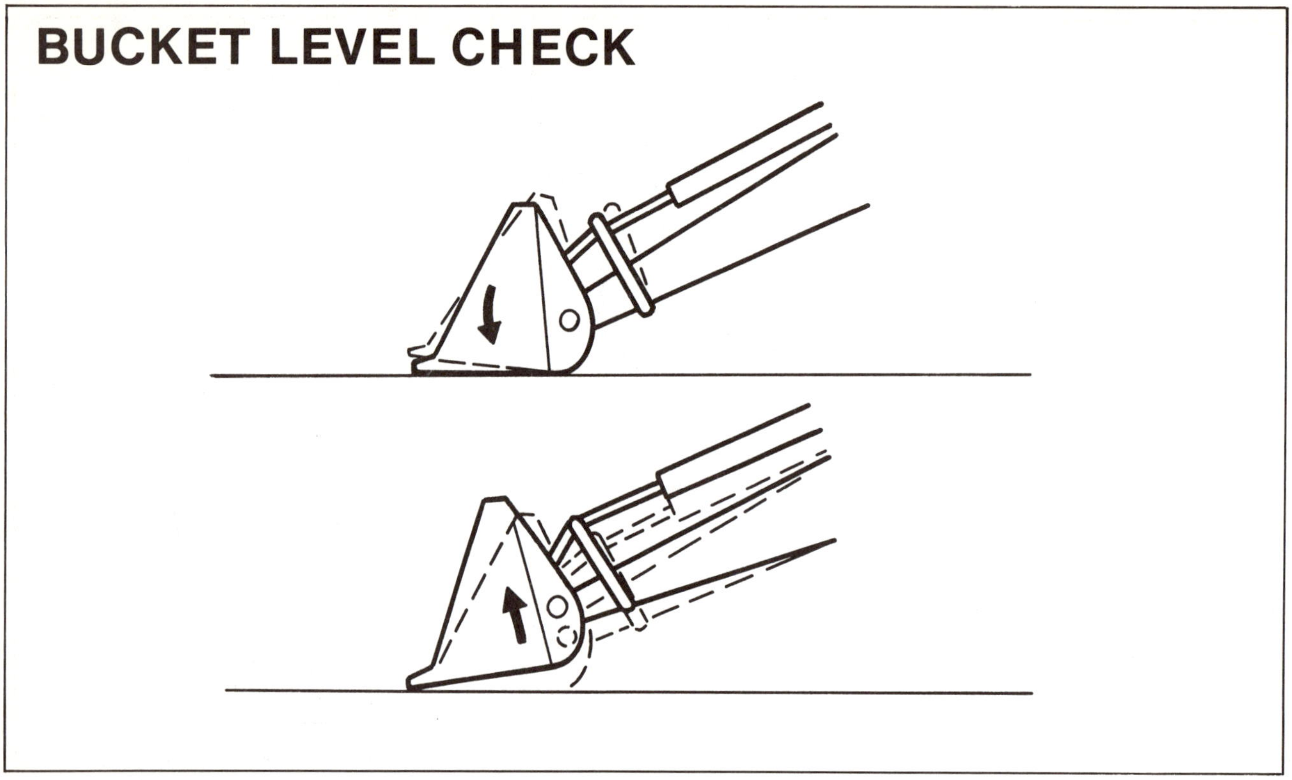

10-2. BUCKET LEVEL CHECK: This procedure can be used to check the accuracy of the bucket level indicators. On machines without an indicator, it can be used to determine the level position of the bucket. The machine *must* be positioned on a *flat* surface. A) Lower the loader lift so it is in the "float" position. Starting with the bucket rolled slightly from the level position, the bucket is rolled down slowly. At the instant the bucket is level, the loader lift arms will begin to raise (B).

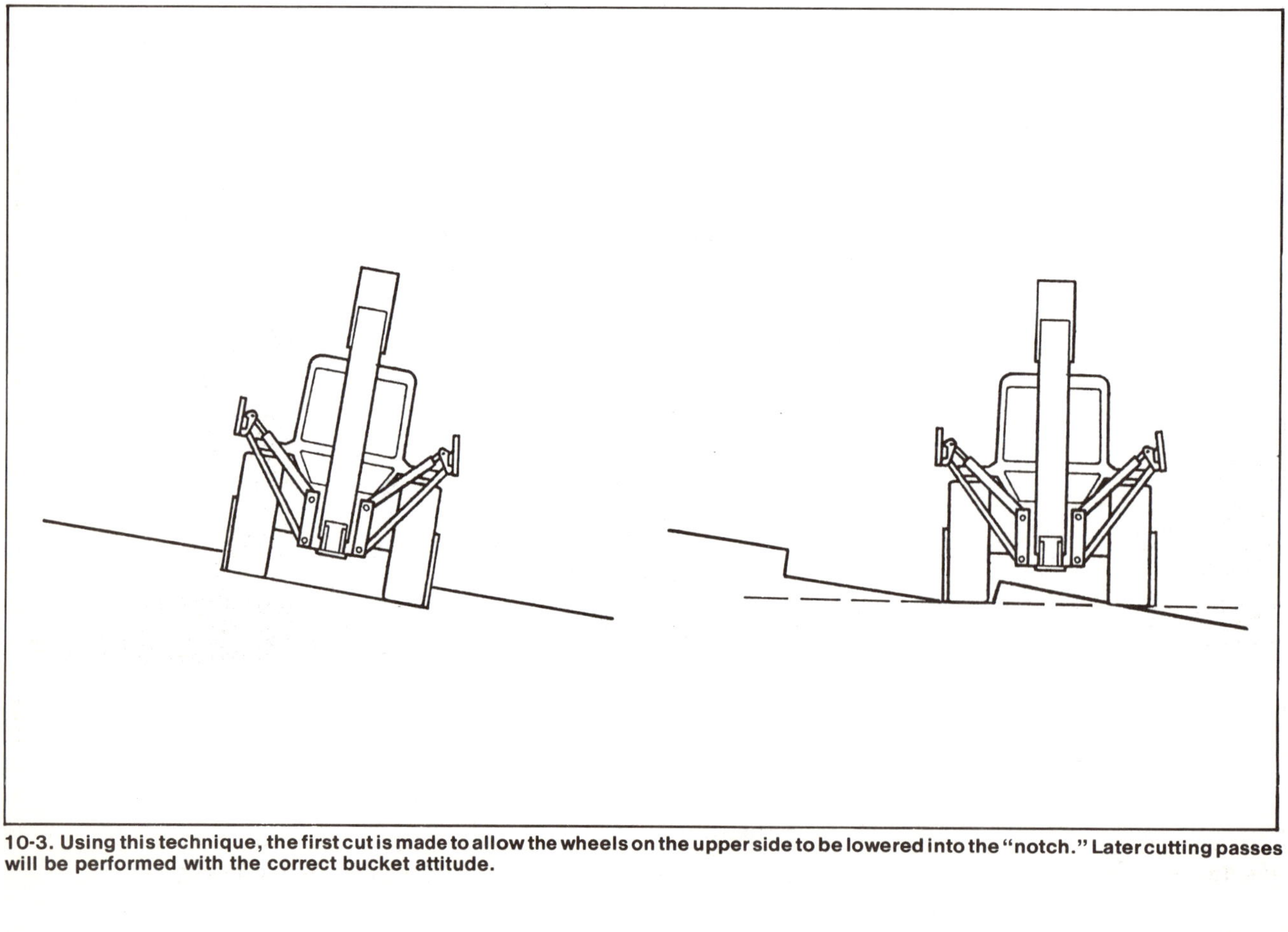

10-3. Using this technique, the first cut is made to allow the wheels on the upper side to be lowered into the "notch." Later cutting passes will be performed with the correct bucket attitude.

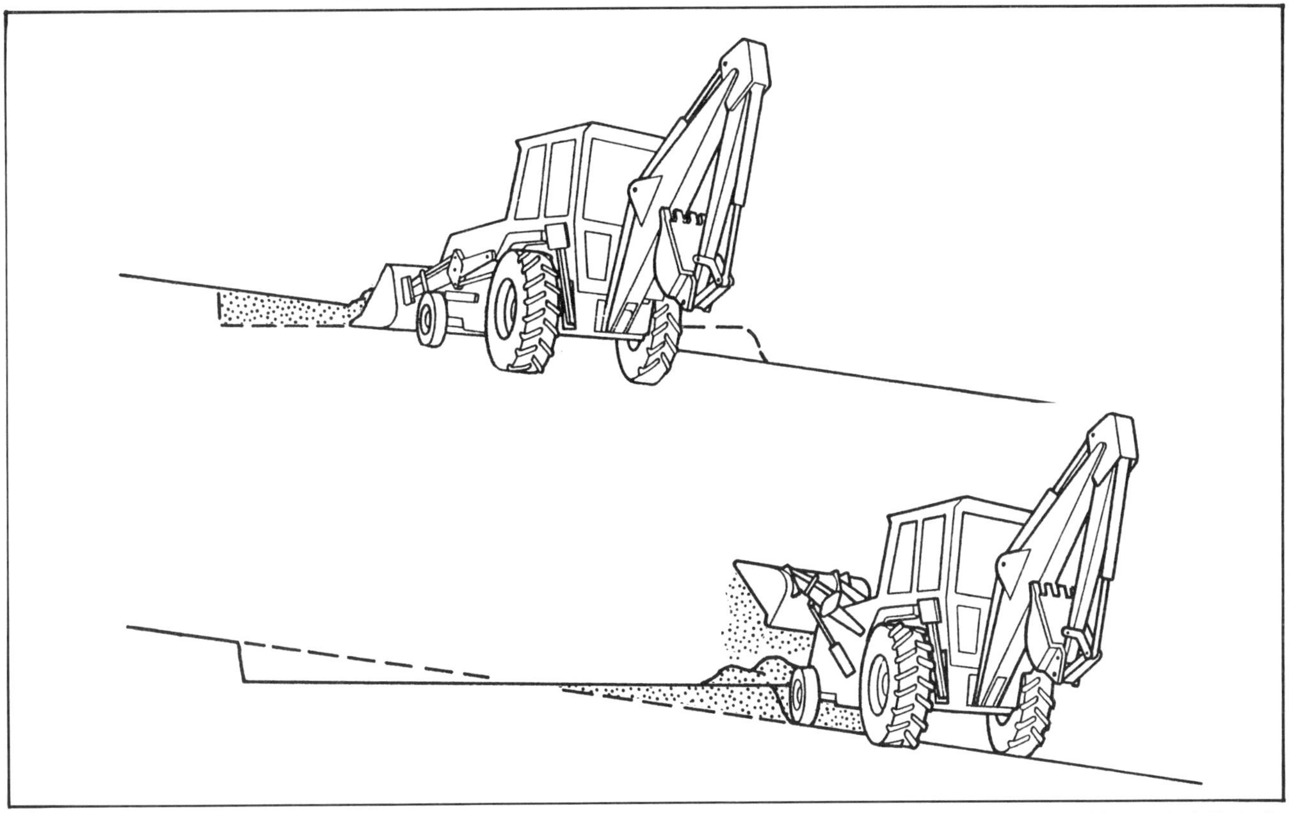

10-4. Using this method, the machine will "attack" the slope at a 45 degree angle. By taking small cuts, and depositing material into the immediate fill area, an area is slowly leveled for the machine to work from. Once the TLB is level, the cutting procedure will progress much more easily.

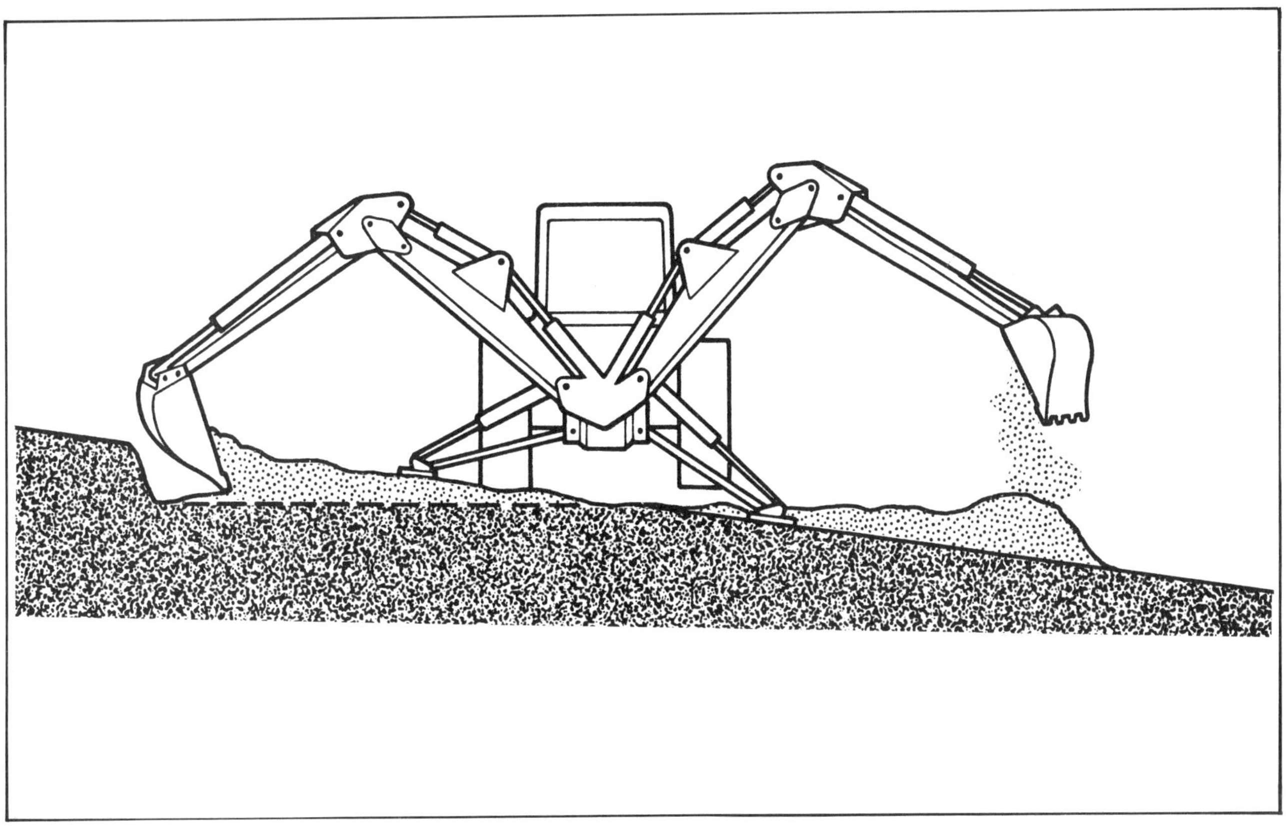

10-5. This is also a very useful technique. The backhoe is used to cut, as well as deposit material into the fill area.

methods that can be used to provide a suitable starting place for the loader to complete the cut. For three examples of how this can be accomplished, see Figs. 10-3, 10-4, and 10-5.

Once the cut is started, and the TLB is supported by an area with the desired **angle of grade,** then the cutting process can be carried out effectively.

As material is removed from the cut area, it is then either deposited in the fill area, or stockpiled. The cut area should be kept graded and polished so that the cut continues as desired. If a particular cutting pass is either too deep or not deep enough, the mistake can be corrected on the next several passes.

If soil in the cut area is of medium hardness, or soft, cutting will be easier than when hard soil conditions exist. The bucket can penetrate soft soil easier, and the operator will have greater control over the depth of the cut. When soil is hard, however, the cutting edge won't penetrate as easily. To overcome this resistance, and better control the bucket and the cutting depth, a series of scraping passes can be made. The scraping is done in a definite pattern so the operator can control the depth of the cut more easily (Fig. 10-6A). This pattern will produce a shallow "trough" (Fig. 10-6B) from which the operator can continue cutting by starting in the center, and then working outward in both directions. If increased cutting penetration is needed, the operator may position the TLB at a slight angle in the trough. With the machine in this position (Fig. 10-7), the corner of the bucket will receive more weight and power.

Another way to enhance the degree of control when cutting in hard areas is to increase the weight on the cutting edge of the bucket. This can be accomplished by filling the loader bucket about one-quarter to one-third full prior to attempting the cut. This added weight will enable the bucket to penetrate harder material, and the operator will also have better control over the depth of the cut.

Filling

In any filling or back-filling operation, the actual filling as well as the compaction, and at least the rough grading, should be done as the job progresses. To link these procedures together into a workable job plan (one that is smoothly, easily, and professionally performed), the fill material must be dispersed in such a way that after compaction, the correct grade will have been achieved.

The closer the operator can come to the right grade the first time, the easier the finish-grading will be. On jobs that take an entire day or more, the operator can constantly correct and polish the cut-and-fill areas, and carry out the compacting and grading requirements at the same time, as he travels around the job site. By keeping this in mind the operator will actually save time in the long run and ease the work performed by others on the job site.

To better understand the filling procedure, let's break it into two parts: **dispersing the material** and **compaction.**

1. **Dispersing the material in the fill area.** Depending on the depth of the fill area, the fill material must be dispersed at a level higher than the desired finished grade, to allow for compaction. The extra amount required is usually about one-third of the fill depth. So, a fill area three feet deep will require about one foot of extra material; if the area is six feet deep, about two extra feet of material will be required; etc.

As the depth of the fill area increases (six feet or more), extra care must be taken as the TLB works near the edge. For increased safety, and to aid in the compaction of the fill, material can be dumped into a pile on top of the fill, and then carefully pushed over. On fills of ten feet or more, extreme caution is advised due to the possibility of a slide caused by poorly compacted material in the middle or lower areas of the pile. As an extra safety measure on deep fill jobs, material should be dispersed in an arc to ensure stable ground for machine support (Fig. 10-8).

It should also be understood that the same principles discussed here apply to the back-filling of trenches and excavations.

Dispersion techniques. Here we will explain the techniques involved in the actual fill procedure, for the entire range of fill area depths recommended for the TLB. The fill area we will examine is the other half of the cutting job described earlier. And the filling process is performed at the same time the cut is done—that is, at the same time cutting is started, the spoil from the cut is transported and the fill is also started.

There are basically two methods of accurately dispersing material: in front of the bucket (going forward), and behind the bucket (going backward). The former approach—dispersing material in front of the bucket—is the most usable of the two, and can be used for all depths of fill.

For shallow fills, the bucket can be dumped quickly, spilling the material in front of the bucket. Then, by moving forward slowly and

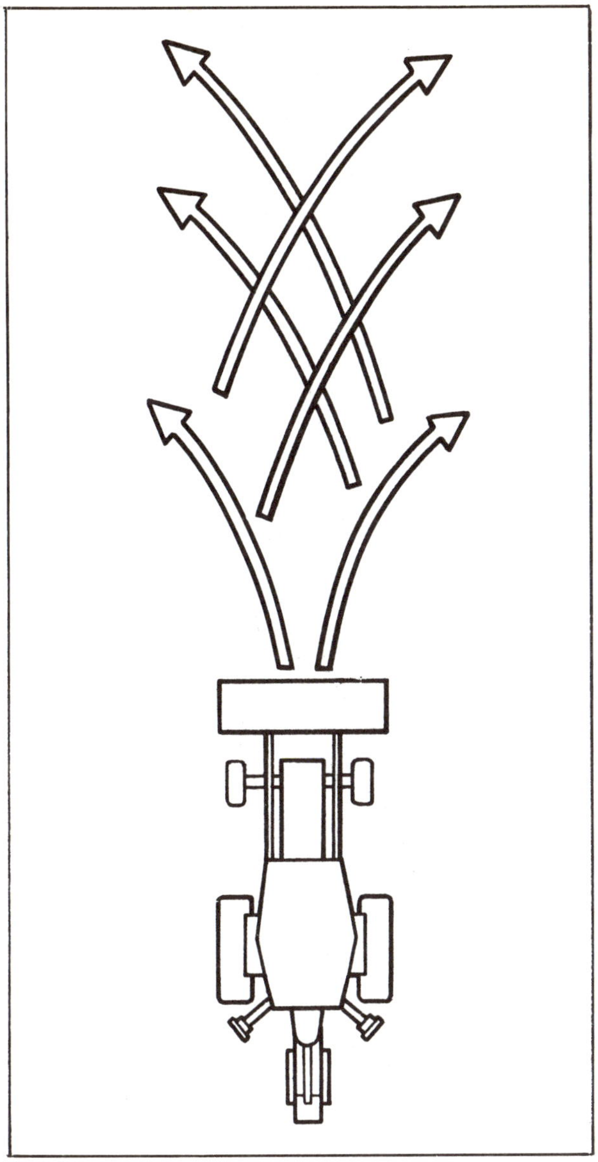

10-6A

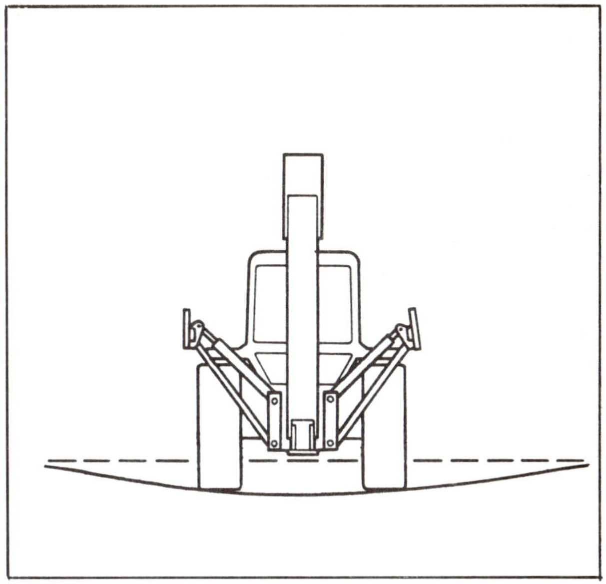

10-6B

FINE CUTTING TECHNIQUES

10-7

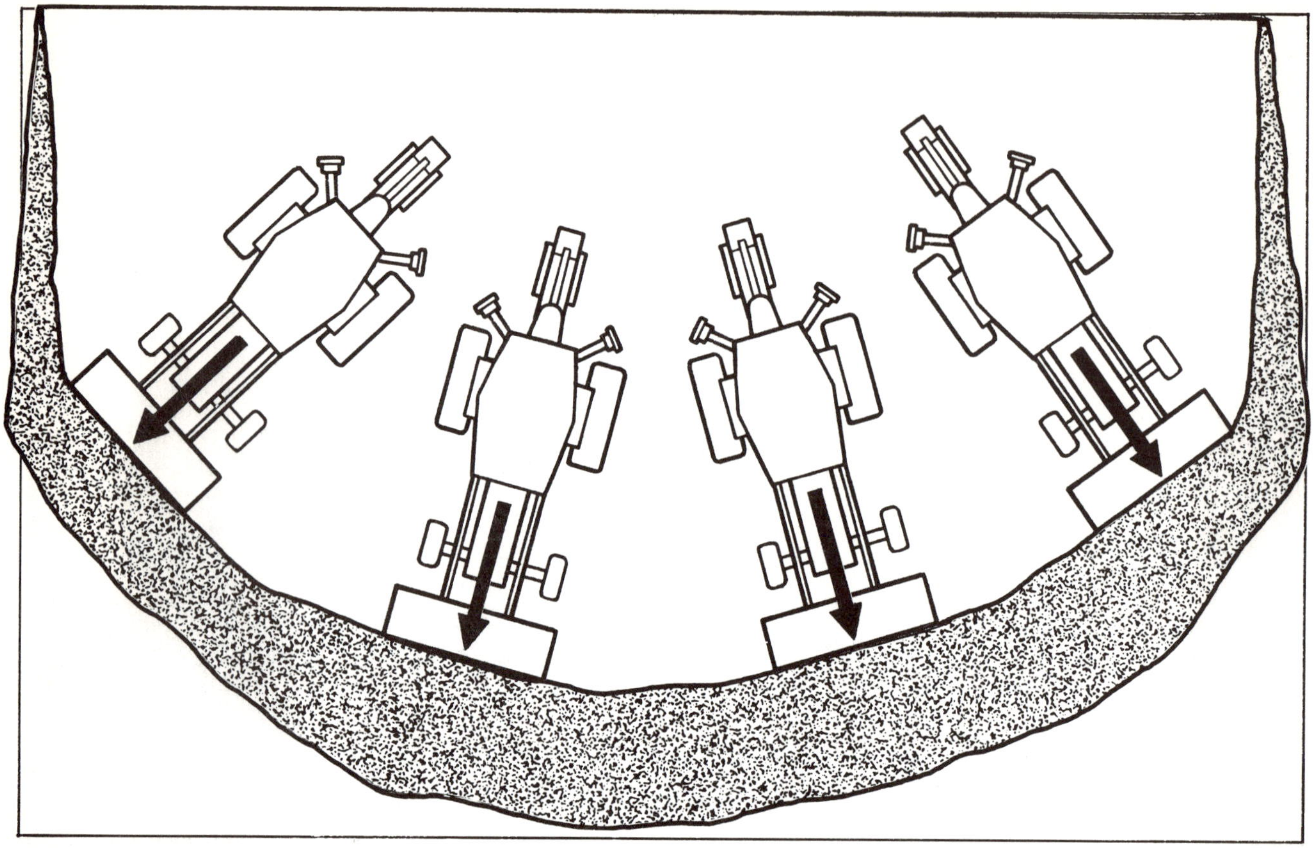

10-8. By approaching and forming the fill slope in this manner, the TLB will always be positioned on solid ground.

carefully, following the desired grade with the lower edge of the bucket, the material is dispersed to the correct depth (Fig. 10-9). As the fill deepens, the material is again dumped on the surface of the fill and pushed forward, piling it high enough to allow for compaction.

Another effective method for shallower and more precise fills is to disperse the material behind the bucket, while traveling backward. Figs. 10-10 through 10-13 illustrate how this is done. Notice that the angle at which the bucket contacts the ground is different in each photo. When a shallow fill is desired, the bucket is positioned to pull the material back and disperse it very gradually. As the amount of material remaining behind the bucket is reduced, the bucket is rolled to allow for even dispersion.

2. **Compaction.** The second phase of the filling procedure is stabilizing the fill material, which is accomplished by compacting it, using the weight of the TLB (Fig. 10-14).

The fill area should be compacted as the fill procedure goes on, and the grade corrected as the work continues. The grading and compaction in this type of situation can usually be accomplished with little delay before returning to the cut area for more material. The operator can effectively compact the fill when the bucket is full. The TLB depicted in the photos, with a loader bucket capacity of one cubic yard, will carry 1,700 to 3,250 pounds when full. The weight of the material will vary depending on its water content. In fact, 2,000 pounds of dry soil can weigh up to 3,000 pounds when wet.

Using a full loader bucket and compacting the area with the TLB's front tires will result in the highest degree of compaction capable of the TLB, in the absence of specialized compaction attachments.

Finish Grading

Finish-grading (or fine grading) involves essentially the same principles as rough grading, except that it requires much more precision. Instead of moving the cut-and-fill material in bulk to accomplish the task, finish-grading requires the cutting and filling of as little as one to four inches of material. **If** the rough grading was done correctly, very little additional cutting or filling should be needed—perhaps only a few inches in some spots. Of course, on any finish grading project, the entire area should be smoothed with the loader to produce a clean,

10-9. When using this method, the bucket is dumped quickly, and the fill depth is adjusted to the correct level by raising or lowering the bucket. In order to achieve accurate results, the TLB must move forward slowly so the operator can make necessary bucket adjustments.

even, textured surface—free of dips, bumps and even tire marks.

Finish grading technique. Because of the different ground conditions that will be encountered, it is difficult to explain any one finish grading technique that can be used in all cases. However, the method which produces the best finish grading results is to position the loader bucket **flat.** By using the bucket at a flat or very nearly flat angle, the cut-and-fill can be adjusted to a very fine degree.

At the start of a finish grading job, the site may contain ruts, bumps or any variety of surface irregularities. As work gets underway, the operator will first eliminate the larger bumps, and then continue smoothing and polishing, using special grading patterns which will produce a perfectly smooth area.

When fine grading is required, **back-dragging** will produce the best results. This is best accomplished with the loader in the **float** position, and the loader bucket tipped to produce a slight cut. This slight cutting action will continue to fine grade the area as back-dragging continues. Fig. 10-15 illustrates the results of this slight cutting action, while Fig. 10-16 shows the results of slight filling, or allowing the material behind the bucket to flow under it.

To achieve the final "polished" results, the operator should vary the direction of the grading passes. This change in direction will effectively smooth the area and leave it with an evenly textured surface (Fig. 10-17).

When ground conditions are about average, with normal soil hardness, the procedures shown in Figs. 10-18 and 10-19 represent the most useful finish grading method:

Fig. 10-18: The loader bucket is loaded about half full, to provide extra weight as well as a "reserve" of fill material. The TLB is first backed up to the far end of the grading area. The loader bucket is then lowered to the ground in the flat position. The operator begins to slowly move forward. With the bucket perfectly flat on the ground, it will not make a cut, nor will it add any fill to the area. To make a cut when desired, the operator may tip the bucket **forward** slightly, and for a fill, he will tip it **up** slightly. In areas where no cut or fill is required, the bucket is returned to the flat position. When the machine has traveled to the farthest area where grading is required, the bucket is raised just high enough to clear the material that has been accumulated, and then moved past it.

Fig. 10-19: Then, this material is **back-**

10-10. The fill material is dumped, and the bucket is moved past the pile. This photo shows the operator applying down pressure to the bucket to prevent it from sliding over the pile. Once the pile begins to move, the loader control is placed in the "float" position.

10-11. As the machine backs up, the bucket is slowly rolled out, more quickly if a deep fill is desired, or less quickly for a shallow fill.

10-12. Continue rolling the bucket as the machine travels. . .

10-13. . . . and the result will be a fill area that will always be tapered into the existing grade, with a smooth working area maintained.

10-14. The operator should always fill ahead of the compaction area to ensure safety.

dragged; first, with the loader lowered all the way, and the bucket tipped up (or back) slightly. This action will trap the accumulated pile behind the bucket. A large pile will offer great resistance, and will require extra weight and a slight cutting action with the bucket. After the pile begins to move, the extra weight of the loader is removed by placing the control in the "float" position.

Backhoe Grading Techniques

Although the grading capabilities of the backhoe are very limited, it can often be used in situations where the loader cannot, or where the loader operation would be unsafe. These situations may arise in a variety of circumstances, such as job sites with poor access; a working area too small for effective loader use; ground conditions too soft to support the TLB's weight, or provide necessary traction for effective loader use; or where slopes or differing elevations do not permit loader use.

One example of how the backhoe may be effectively used for grading work is when the operator must "scratch out" a driveway. This common procedure is necessary whenever a dirt or gravel driveway is to be replaced with concrete. Dirt will usually have to be removed three to five inches below the existing grade. Fig. 10-20 illustrates the procedure.

In other situations, the **swing** can be used for grading small areas that are either flat or sloped. When using the swing of the hoe to grade, it is important to utilize the backhoe's hydraulics to provide the necessary force, and not the momentum of swinging, to push heavier loads.

The point here is that the swing motion of the hoe can move the bucket faster than it should be for this type of operation, and this should be avoided. The hazard presented by swinging too fast is that a hidden obstruction could suddenly stop the bucket, and severe damage to the structural members of the hoe could result. **Never swing the hoe deliberately into a solid object,** such as a wall or stump, or anything that could suddenly stop the boom and crowd assembly. Instead, the swing should be operated slowly and carefully during this type of procedure.

10-15. Back-dragging with the bucket slightly tipped up will produce a cutting action. The cutting abilities of this technique can be increased by adding weight with down pressure, or by adding material to the bucket.

10-16. When a fill is desired, the bucket is tipped down slightly. The fill material behind the bucket will flow evenly under it.

10-17. After back-dragging the entire area in one direction, the direction of the passes can be changed 60 to 90 degrees to put the finishing touches on the area. Finish grading using this particular grading pattern will produce excellent results.

The stabilizers can often be used to tilt the machine slightly in one direction or the other to adjust the grading results. To produce the optimum results, the bucket must be manipulated in such a way that it will produce a flat area, rather than a rounded one, as it contacts the ground (Fig. 10-21).

Compaction

Principles of compaction. Depending on many factors—local weather, water tables, and the presence or absence of rock, sand or gravel—soil may be in any consistency when first encountered. It may be dry or wet, hard or soft, or any condition in between; it may fall apart easily, or be quite difficult to break up. In any case, however, once the soil has been dug up and worked around the job site by the TLB or other earth-moving equipment, it will be fairly loose. For this reason, filling or back-filling will often require compaction of the material.

The compacting procedure could involve anything from light wheel-rolling to the use of specialized compaction equipment or attachments. In some areas, strict testing of soil density is required prior to and during construction, to minimize the risk of future subsidence or other ground movement. In these situations, soil engineers will work with the TLB operator, testing the percentage of compaction as work progresses. During these operations, several principles of compaction must be followed to achieve the desired results. Essentially, compaction is accomplished by applying great weight or impacts on the area to be compacted. These compaction forces compress the soil downward and outward at an angle, as shown in Fig. 10-22.

For the actual compaction of the soil to occur, two things must happen: excess air and water must be forced out of the soil and out of the fill area, and the individual particles of soil must be compressed together very tightly.

Excess air and water removal. To achieve the right soil density through compaction, soil must be placed in layers thin enough to permit the expulsion of air and water. Some soils, depending on their content, may be placed in thicker layers, or "lifts," than others. This is because some soil types allow greater movement of air and water than others. For example, sandy

10-18. When fine grading an area, the ideal method is to cut off the high spots on the forward pass. . .

10-19. . . . and use the accumulated material to fill the low areas on the backward pass.

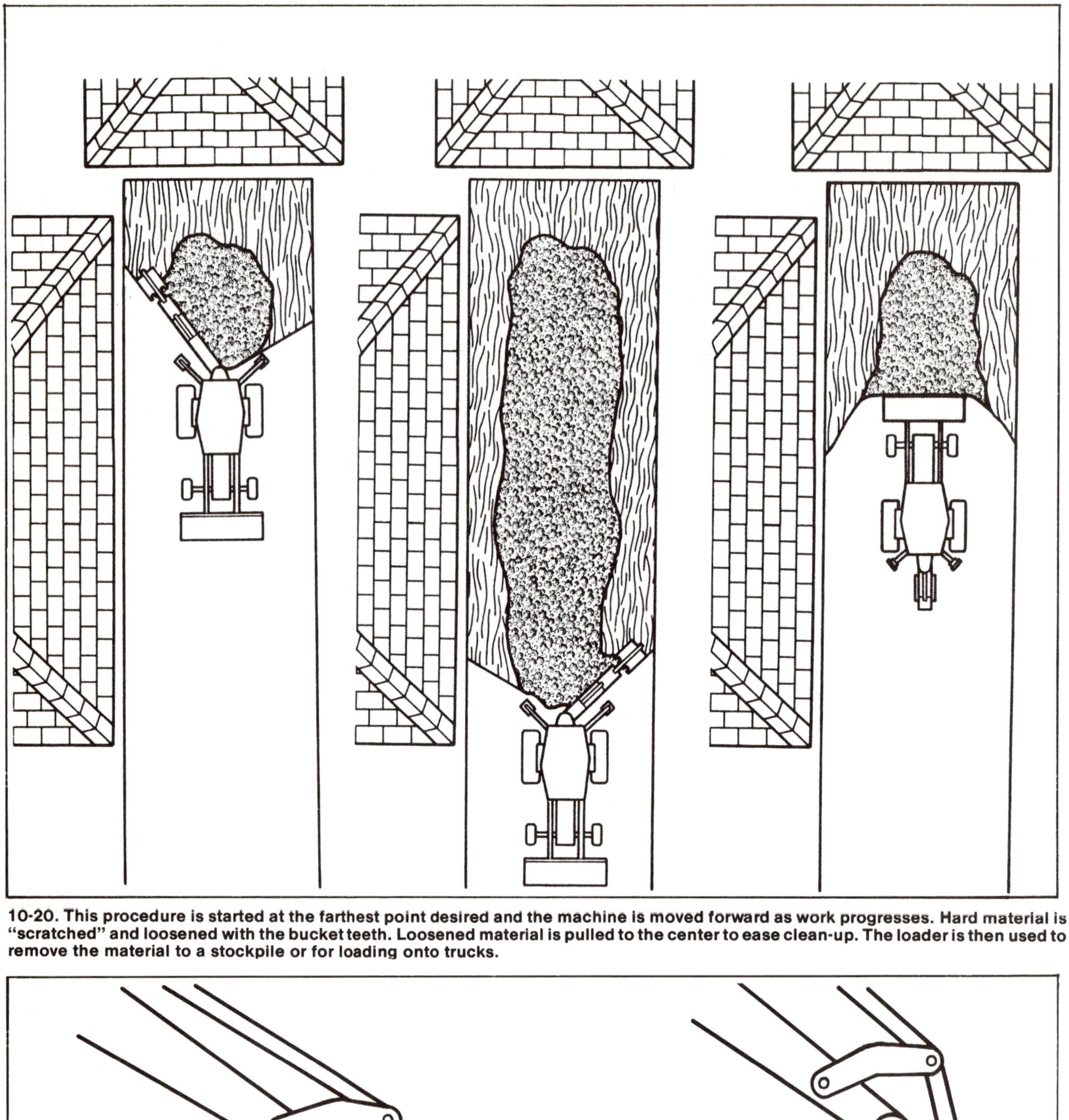

10-20. This procedure is started at the farthest point desired and the machine is moved forward as work progresses. Hard material is "scratched" and loosened with the bucket teeth. Loosened material is pulled to the center to ease clean-up. The loader is then used to remove the material to a stockpile or for loading onto trucks.

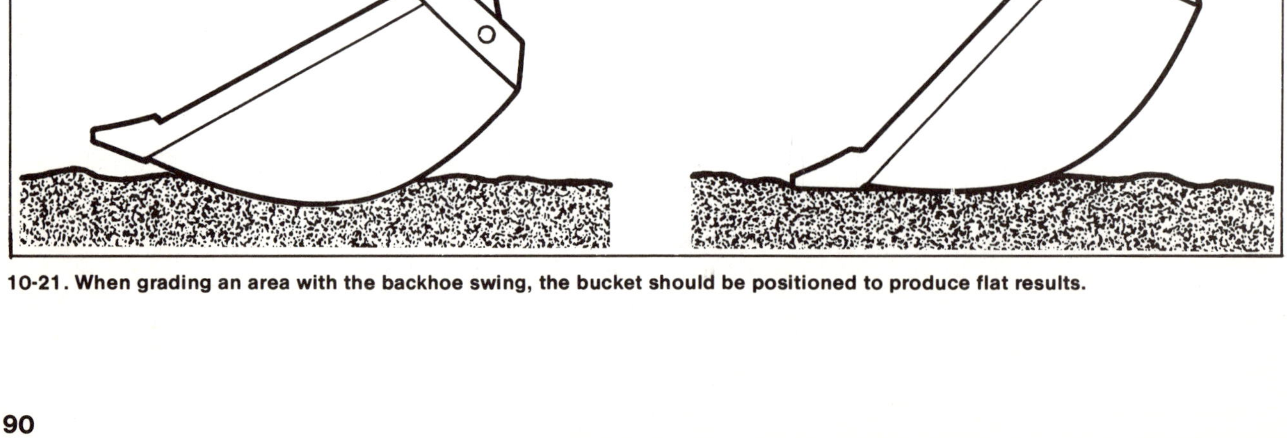

10-21. When grading an area with the backhoe swing, the bucket should be positioned to produce flat results.

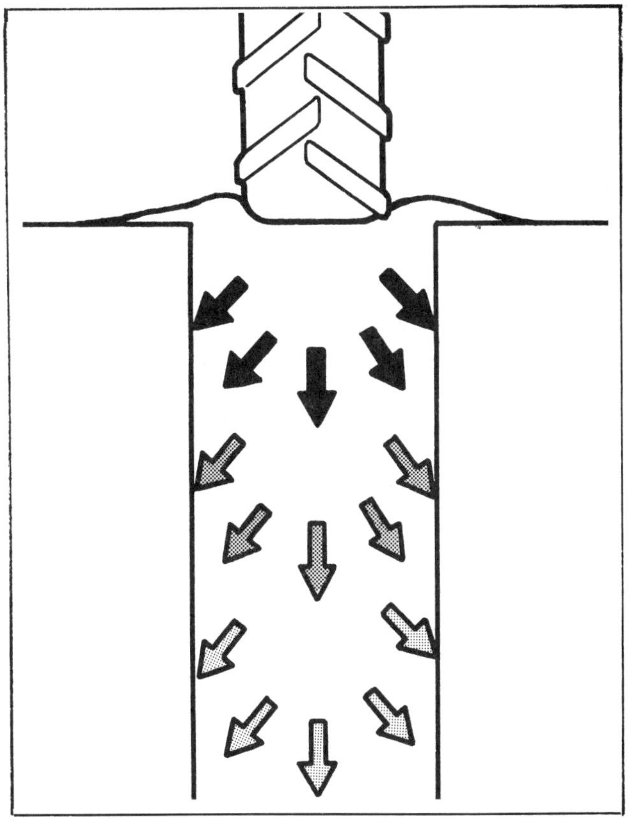

10-22. As the machine drives over the area, the forces not only radiate straight down, but also at an angle into the sides of the excavation. Compaction is greatest near the surface, and forces are reduced as depth increases.

soil allows air and water to pass through fairly easily, so it can be layered in thicker lifts than clay, which does not easily allow air and water to flow through.

To achieve the highest degree of compaction, soil is often placed in lifts of about twelve inches. When compaction is not a critical factor, lifts may be increased in thickness to eighteen or twenty-four inches, to speed the filling procedure.

Compression. Viewed on a very small scale, compaction serves to compress individual soil particles together. When soil is dry, this compression is resisted due to friction between the particles. However, this friction can be effectively reduced by providing lubrication to the soil, in the form of water. **Compaction cannot be done effectively with dry soil.** Once the soil becomes wet, the individual particles will come together more closely, and the area will begin to sink.

On the other hand, the soil shouldn't be oversaturated. Too much water can make the soil muddy or mucky, and compaction will be impossible. Also, too much water can actually prevent compression when the excess water cannot escape into the surrounding soil.

In order to achieve the optimum water content, soil should be moistened evenly, with enough water to dampen it, but not so much that it will become muddy or mucky. Since water content is critical, it is a good idea to have a worker moisten the soil as it is filled or back-filled into the trench. When soil is very dry, it will be necessary to push or dump it into the fill area very slowly, to ensure even and complete moistening.

Jetting. This is another useful method of using water as a compaction tool. It involves the injection of great amounts of water into the area to be compacted, using a water hose equipped with a solid pipe on the end (Fig. 10-23). By injecting water in this manner, it will actually perform most of the "work" of compaction, the way it usually occurs in nature. This technique is more useful in sandy or gravelly soils than clay or other soils that do not allow the water to escape easily. It can also be effective when applying quick road patches, as it provides stability to the area.

Figs. 10-24, 10-25 and 10-26 depict how jetting can be applied most effectively under several different soil conditions.

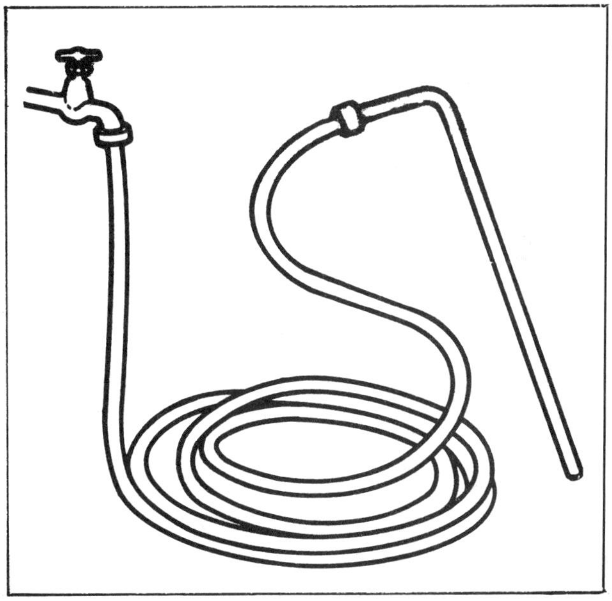

10-23

Wheel Rolling

Although wheel rolling may seem to consist simply of running over an area with the TLB's wheels, it must be done very methodically. Filling and back-filling are common TLB procedures, and in many cases, compaction will also be re-

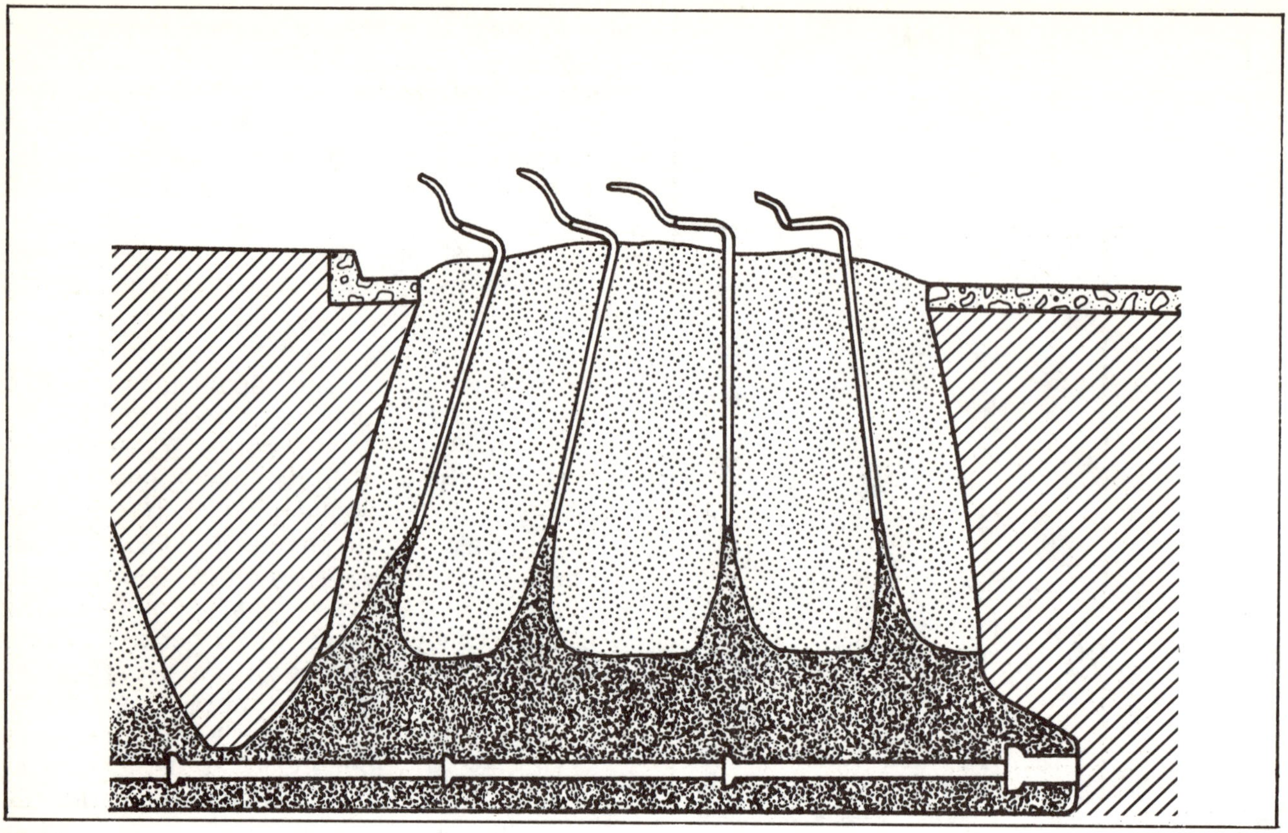

10-24

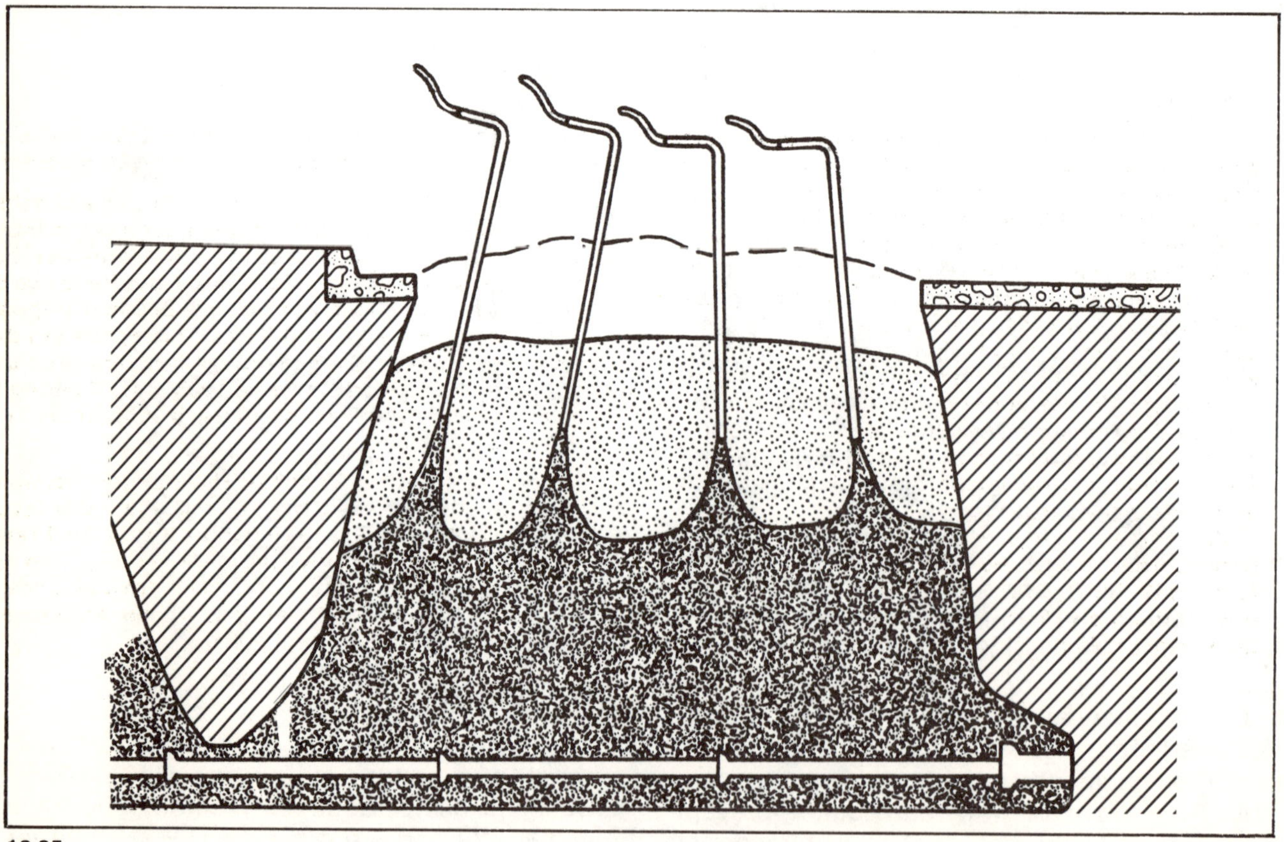

10-25

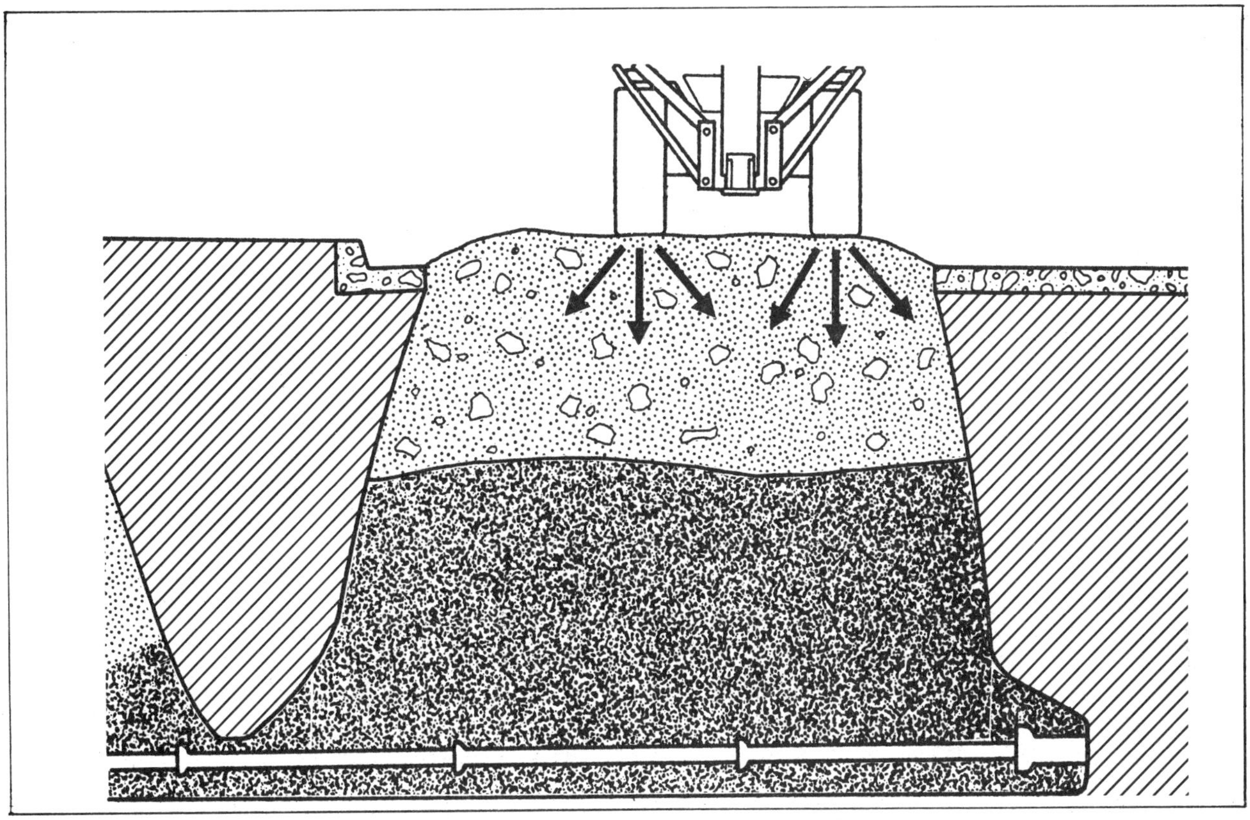

10-26

JETTING

Fig. 10-24. Here a trench is back-filled to the existing grade with loose spoil. Water is introduced at the lower end of the trench; as that area becomes flooded, the jet is moved to the other sections of the trench.

As the trench is compacted from the bottom, the back-filled material will gradually sink. The water level is then raised by bringing up the end of the jet. Again, the jet is moved around to various sections of the trench so that the lower levels are thoroughly saturated. If the water seeps quickly into the material and no puddles accumulate on the surface, the fill material can be saturated to within twelve inches of the existing grade. If the water does not soak in easily, then the saturation level should not be raised higher than twenty-four inches from the existing grade.

Fig. 10-25. By keeping the top layer of the fill area from being saturated, the area can then be safely wheel rolled with the TLB without becoming unstable. However, if the ground does not accept the water easily, then wheel rolling will force it to the surface, and the area will become mucky and unstable. If the material accepts water easily, this usually won't be a problem.

Fig. 10-26. Here we see an area with clay soil being jetted. The most important consideration when introducing water into this type of soil is to remember that the water will not be easily absorbed into the sides of the excavation. Water will instead become trapped within the fill material, and if weight is applied to the area, it will rise to the surface, making the area unstable for the support of wheeled vehicles. If jetting is done correctly, however, the area will support the TLB without sinking in.

In order to achieve stability in these situations, water must be injected in such a way that the lower levels will *not* be flooded, and the top three feet of the fill area should be moistened only if the soil is very dry. This way the dry section near the surface will provide a "cap" on the moistened level underneath, and will provide adequate support for wheel rolling.

10-27

10-28

quired, as we have noted. Wheel rolling is the most common technique for accomplishing the necessary compaction.

In order to achieve the **highest degree** of compaction possible, the loader bucket should be filled with heavy material. With the bucket loaded, the operator then rolls the front wheels over the entire fill area. This procedure is then repeated two, three, or even four times, until the area is completely stabilized. It is important to note that when the loader bucket is loaded, the front wheels will achieve a greater degree of compaction than the rear wheels, even though more of the machine's weight is concentrated over the rear wheels.

It should also be pointed out that the operator must always exercise great caution when entering and traveling over the fill area. The safest approach is to load the loader bucket about half full, and then **slowly** move the front tires into the fill. Watch the fill material carefully as the tires enter the area; if they begin to sink deeply into the fill material (Fig. 10-27), the bucket can be lowered as the operator backs the machine up. This will fill the depression caused by the front tires, and prepare the area for the next wheel rolling cycle.

This fill-compact-fill process is continued until the area is stabilized enough to hold the weight of the TLB. Then, the loader bucket may be loaded as full as possible, for added weight. (Remember that wet material is much heavier than dry, and therefore will achieve greater soil compaction.). Then the TLB is repeatedly driven forward and backward over the area, until the desired results are achieved (Fig. 10-28). Once compaction is completed, the area is finish graded, as seen in Fig. 10-17 (see ***Finish Grading***).

Compaction Attachments

Because soil compaction is such an important element in structural stability, it is usually given much attention during the construction process. The contractor has several options available to him to ensure proper compaction; the TLB, equipped with any one of several compaction attachments, is one of these options. There are several types of attachments available, and some are more effective than others.

The most common type of attachment is basically a large "vibrator," consisting of a hydraulic motor mechanically attached to an eccentric weight. When actuated, the spinning weight causes the attachment to vibrate very forcefully.

Another type in common use is a "hoe-ram" or hydraulically operated jackhammer attachment. A plate is attached instead of the hammer point, and the force of the repeated blows compacts the soil very effectively.

These two types of compactors aren't capable of compressing large areas efficiently, but are useful in a variety of potential problem situations. For example, when plumbing and electrical contractors are required to compact trenches or other excavations, a TLB with a compaction attachment is an excellent choice. With labor costs skyrocketing, and with the manpower required for manual compaction, the TLB also becomes an economical choice.

In addition, when deep excavations require compaction, the TLB may be the only choice from a safety standpoint. The vibration caused by most compaction methods makes it unacceptable to put laborers into the excavation. The TLB, however, can work from the top, in a safe set-up, and be out of danger in case a cave-in occurs.

11. BACKHOE TECHNIQUES

Up until now we have explored various aspects of TLB operation and issues relating to operation, such as maintenance and safety. In the next two chapters, however, we will dig into the specifics of the TLB operator's day-to-day excavating work.

The TLB is designed to perform a number of different tasks, and is capable of some applications even the manufacturers probably didn't think of. The bulk of the work to be performed with the TLB, however, is basic earth-moving, and this is accomplished with one or the other of the two buckets. In this chapter we will discuss techniques involved in the use of the **backhoe** as an earth excavating and transporting tool, and how the operator can master its use in the completion of whatever job he is asked to perform.

Purpose of the Excavation

Because of the wide range of tasks which the TLB is able to perform, there is also a wide range of different job requirements the operator will need to be familiar with. When any digging job is begun, the most obvious task is the movement of the earth that will be required. Although this may seem to be the most important aspect of a digging job, it is often the part of the procedure that will require the least amount of attention. There are many other aspects that are more important to achieving safe, professional results, depending on the type of excavation, and we will discuss them here.

Movement of the earth. Even the simple task of removing the amount of earth necessary for an excavation includes other important, but less obvious, considerations. In fact, digging is only half of the process—the other half is **correct placement of spoil.** While this may not be a vital element of completing an excavation, it is critical to the completion of the job as a whole.

If, for example, an excavation must be back-filled after a section of pipe has been laid, then the spoil should be placed so it can be easily back-filled using the loader. Or, if spoil is to be removed from an area and stockpiled elsewhere, it must be placed in such a way that picking it up with the loader will be easy. If, on the other hand, the excavation will be a working area for other workers, the spoil must be placed far enough away from the excavation so the weight of the pile won't cause a cave-in.

These are all examples of how the operator must think about the entire scope of a job, from beginning to end. Adequate planning can save the operator time and work, and the job will generally proceed more easily, and with fewer problems.

Safety of other workers. The safety of any procedure is, of course, the most important aspect of any TLB job, and the operator is responsible for many of the safety aspects of a job. Leaving aside for a moment many of the other ways the operator is responsible for safety on the job, let's examine how he can provide a safer working area in and around the excavation.

Some of the safety measures the operator can take include: sloping the sides of the excavation when necessary; smoothing the sloped area and dislodging any exposed rocks, chunks of dirt or any other materials which may present a hazard; cleaning the top of the excavation of loose spoil, rocks or other materials that could fall back in; and monitoring the soil conditions, warning other workers of possibly dangerous situations.

Precise excavation. As explained in the chapter on footings, precision digging requires that the excavation be dug in an exact location, to an exact elevation, and usually with a flat, level bottom. This kind of precision is usually only necessary for footings.

Level final grade. This is a requirement common to excavations and trenches. The finished bottoms of excavations for septic tanks, electrical or telephone vaults, leach fields, etc., must always be level. In addition, trenches for footings, leach lines, and many types of pipelines must also usually have level bottoms.

Required depth. Utility lines are commonly buried to a predetermined depth to make locating them easier. Depending on local regulations, company policies, job standards or requirements called for in the blueprints, the only trench requirement may be one of depth, as measured from the ground surface. For example, electrical mains may be required to be four feet deep, while a gas main might only be required to go down thirty inches. In such cases, the trench will be dug to the desired depth, and will follow the contour of the surface—even up or down slopes, if they are present.

Gravity flow trench. This is a common requirement for sewer lines, drain pipes and drainage ditches. Some jobs are very exacting, while others will have plenty of "fall" and will be less exacting.

By studying these different job situations, it is easy to see that all excavations have a definite purpose, and a set of different requirements to achieve that purpose.

In order to put all of this knowledge to use, the operator must ask certain questions before starting work:

—What is the purpose of the excavation—what will it be used for?

—Will the excavation later be a working area for other personnel, and if so, what can be done to ensure their safety?

—Will the excavation require precision digging? What tools and assistance will be needed in order to achieve the desired results?

—Will the trench require a level bottom? What procedure should be followed to achieve this result?

—If the only trench requirement is that it be dug to a certain depth, what tools or procedures can be used to accomplish this?

—Will the trench serve as a drainage ditch, or be used for a sewer line or other gravity flow application? What digging procedure can be followed to get the desired results?

—How much material must be removed? Will there be enough room for a stockpile? Will the spoil be used for back-fill soon? If not, will it have to be stockpiled elsewhere?

Extendable dipperstick procedure. The extendable dipperstick now featured on newer TLB models can ease the digging process greatly, and can speed the work of digging by reducing the number of set-ups necessary. There are several ways the work of the TLB is affected when the extendable dipperstick is used:

—The **reach** of the machine is increased. This means the TLB can not only dig **deeper** and **farther,** but the **stockpiling range** is also increased, which is often very helpful.

—When the unit is extended, the **precision** capabilities of the TLB and the operator are **decreased.**

—As the stick is lengthened, the **digging power** of the machine is **decreased.** The farther it is extended, the less power the machine produces at the bucket.

—When the unit is extended, the **stability** of the machine is **lessened,** especially when swinging the load to the side.

It is apparent that while the extendable dipperstick can be a useful feature, it does have some drawbacks. In order to use it most effectively and safely, it should be positioned in the **shortest position possible.** Of course, this doesn't mean the operator should avoid extending it, but he should extend it only as far as absolutely necessary to perform the task at hand.

The digging procedure, when using a TLB equipped with an extendable dipperstick, is very similar to that of a standard hoe. In order to limit the repeated extending and retracting of the unit, steps 2 through 5 in the general trenching procedure are shown in Figs. 11-3 and 11-4.

As the TLB is moved forward for the next set-up in the trenching procedure, the dipperstick should be retracted to its shortest position, as the operator checks the set-up. Then, it is extended outward about half-way, and the **surface cut** and **bulk removal** steps are carried out at the far end of the trench section (Stage 1). After the bulk of the material is removed from this section, the unit is retracted and steps 2 and 3 of the trenching procedure are repeated at the **near end of the trench** (Stage 2).

At that point, steps 2 and 3 are complete, leav-

General Trenching Procedure

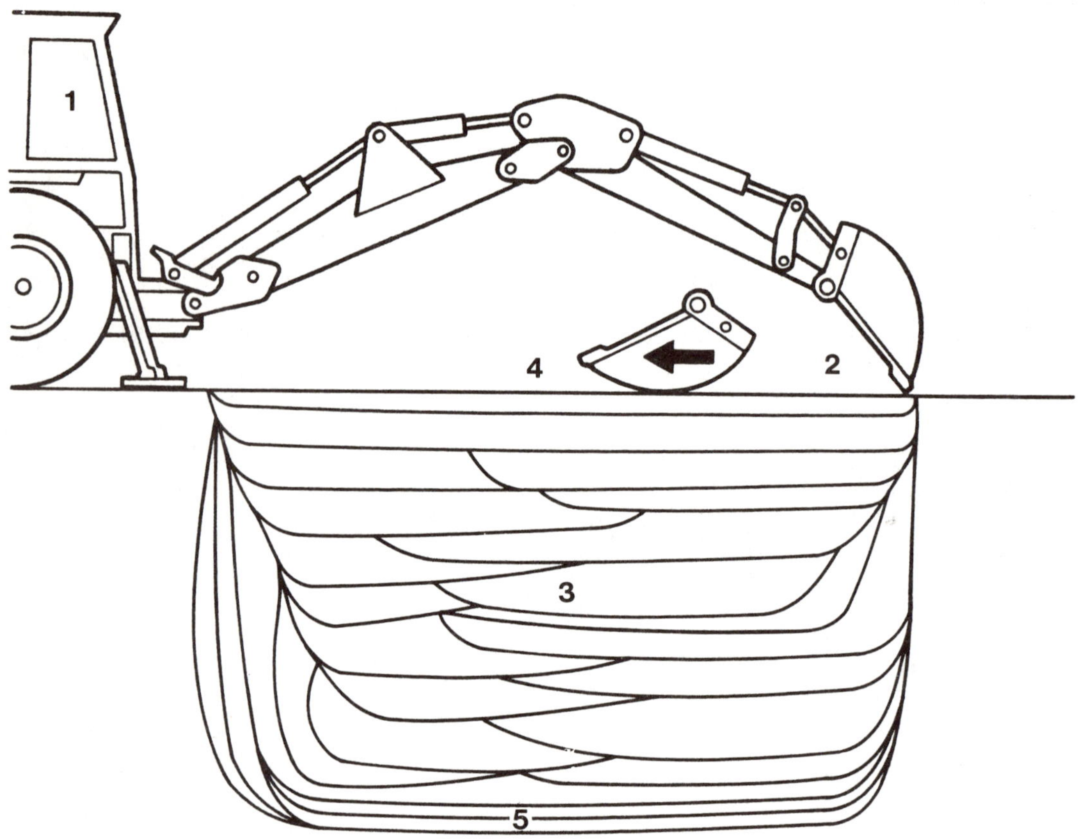

Fig. 11-1

1. **CHECK THE SET-UP** for safety and alignment with the trench.

2. **SURFACE CUT.** The first cut of each section should be shallow, with the operator concentrating on the bucket, making sure the cut is in the exact location desired. This cut will guide the bucket during the rest of the excavation process, so by cutting the first pass in exactly the correct place, the remaining passes will also be correct.

3. **BULK SPOIL REMOVAL.** Dig the trench down to within six inches of the desired grade. Depending on the possibility of utilities in the area, this digging may proceed as quickly as possible, or may demand that digging proceed slowly, with great caution.

4. **PREPARE THE EXCAVATION AREA.** At this point, the operator should clean the sides, or "sidewalks" of the trench. By sliding the bucket flat on the surface area next to the trench, the accumulated spoil will either be pushed back away from the trench, or fall back into it, to be removed in the final grading. Also, any rocks, concrete, or any object which may pose a hazard to other workers should be cleared from the area.

Rocks exposed in the sides of the trench should also be checked and dislodged and removed if they are not securely in place. Any other tasks that need to be done before moving the machine forward to the next section should be taken care of at this time. The area is then ready for final grading of the trench bottom.

5. **FINE GRADING THE TRENCH BOTTOM.** The operator will now make a series of long, grading passes on the bottom of the trench, and also remove any spoil pushed back into the trench during step 4, along with the remaining few inches of dirt. As the trench bottom nears the desired grade, the operator and/or the workers assisting him will check the grade with a measuring tape, builder's level, or whatever grade-checking method is being used for the excavation procedure. After getting the go-ahead signal from the grade-checkers, the operator makes the necessary remaining cuts until the trench reaches the desired grade. He will also remove any accumulated material in the near end of the trench, and make more cuts if necessary to complete the trench as close to the machine as possible.

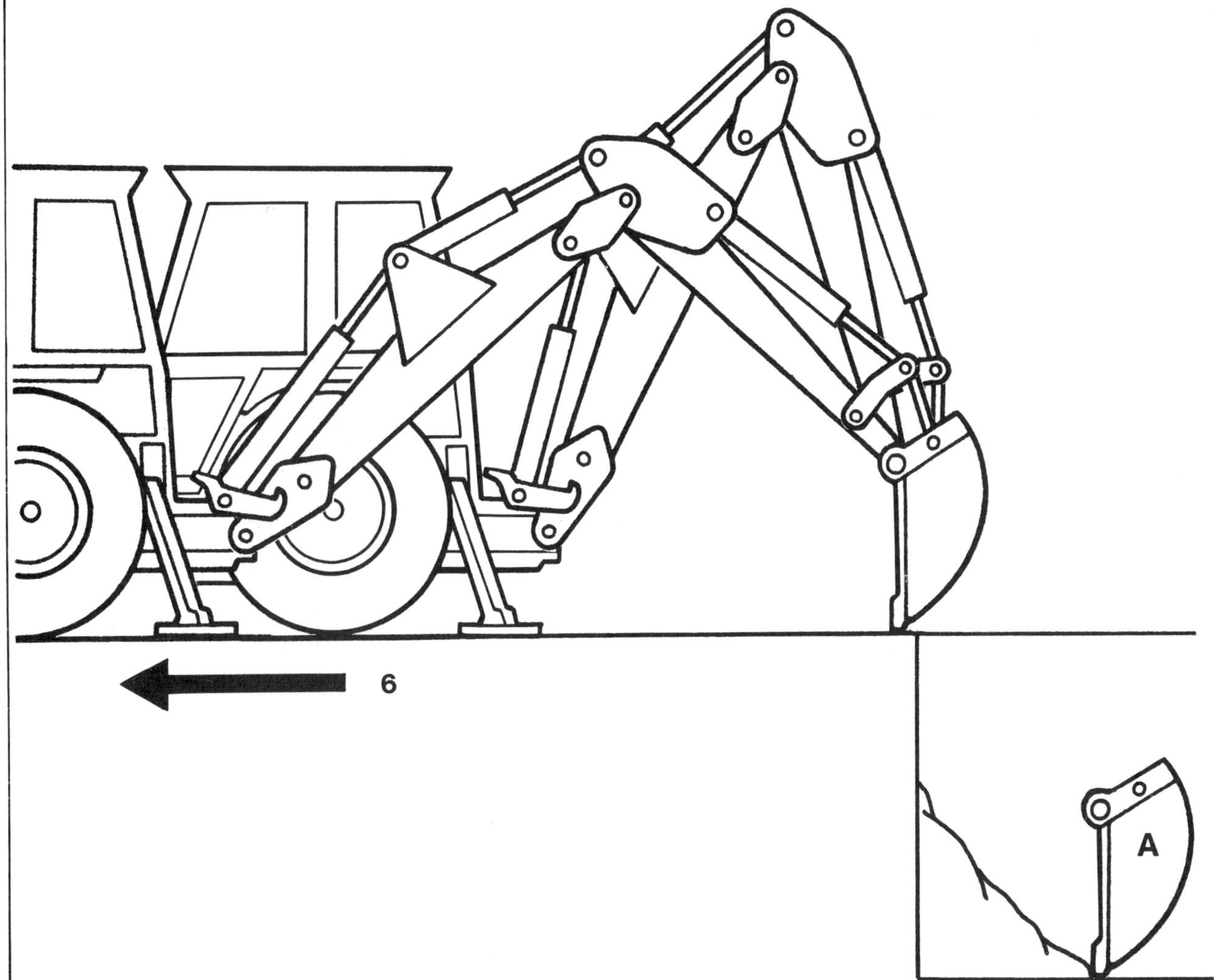

11-2

Fig. 11-2

6. **MOVE TO THE NEXT SET-UP.** At this point, the TLB is moved forward to the next trench section, either by using the drive wheels, or by propelling the machine with the hydraulic power of the backhoe. If the latter method is used, the bucket is first drawn in close to the machine, and lowered until the teeth penetrate into the soil and the machine is held firmly in place. If the tripod set-up is being used, the loader bucket is then lifted off the ground slightly, as well as the stabilizers. Then check the direction of the front tires and turn them in the desired direction if necessary. Check the area for safety aspects, and warn workers in the area to stand back from the TLB.

Now, by using the crowd and the boom, the machine is slowly and carefully propelled forward. The operator should check the bucket to make sure it doesn't raise up out of the soil, which could cause him to lose control of the machine. The point of this procedure is to correctly position the TLB for the next set-up, and therefore, the distance that the machine is moved forward is very important. The machine should be moved as far as possible while still being able to reach the farthest point in the trench where removal of dirt is necessary (A).

The operator should also understand that the deeper the excavation, the shorter each section will be. This is because the bucket's "reach" behind the machine is reduced as it is lowered deeper, and the length of the "flat bottom" the machine is able to dig will be less as well.

As the TLB reaches the next set-up, the entire six-step trenching procedure is repeated, beginning with step 1, the set-up check.

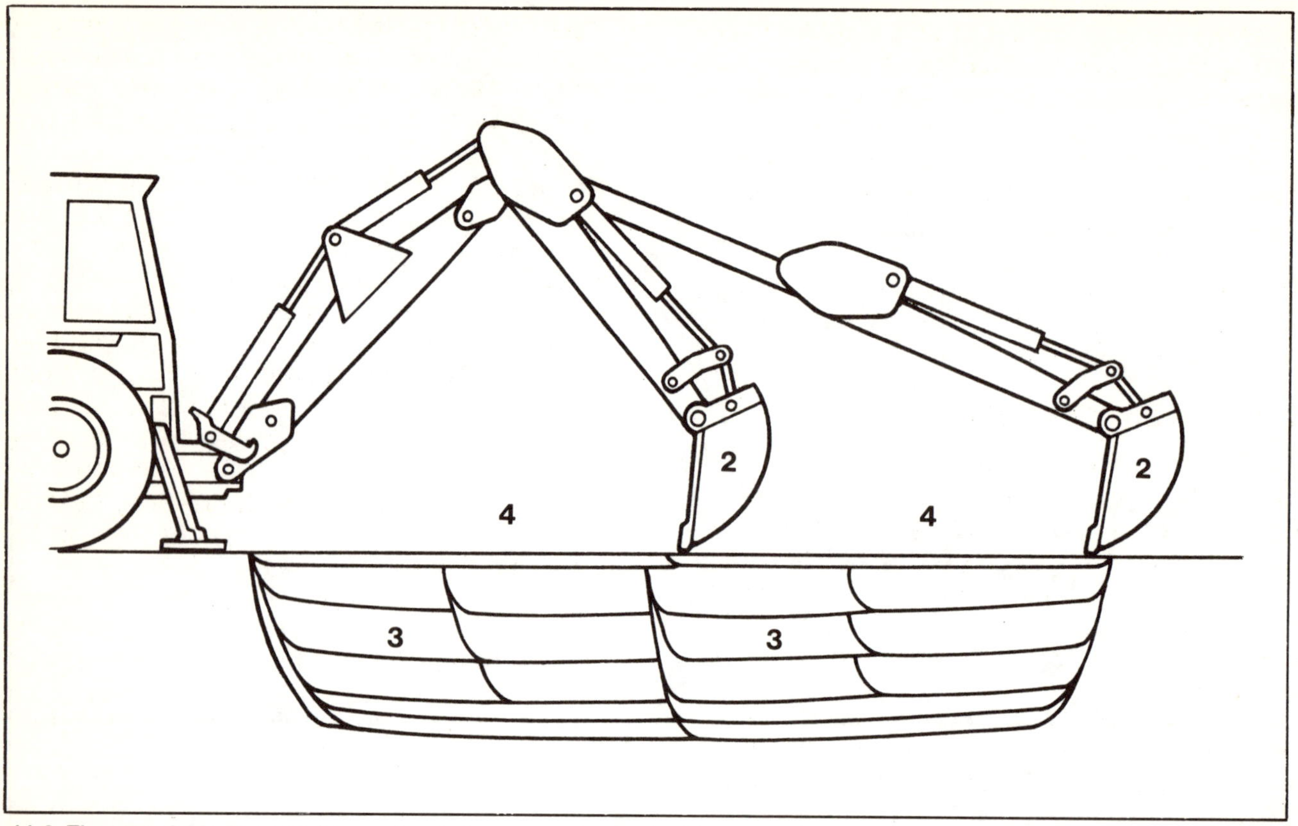

11-3. The extendable dipperstick procedure is the same as the standard hoe procedure, except it is divided into two stages, the first with the stick extended, and the second with it retracted. The numbers refer to the same steps as the standard hoe procedure.

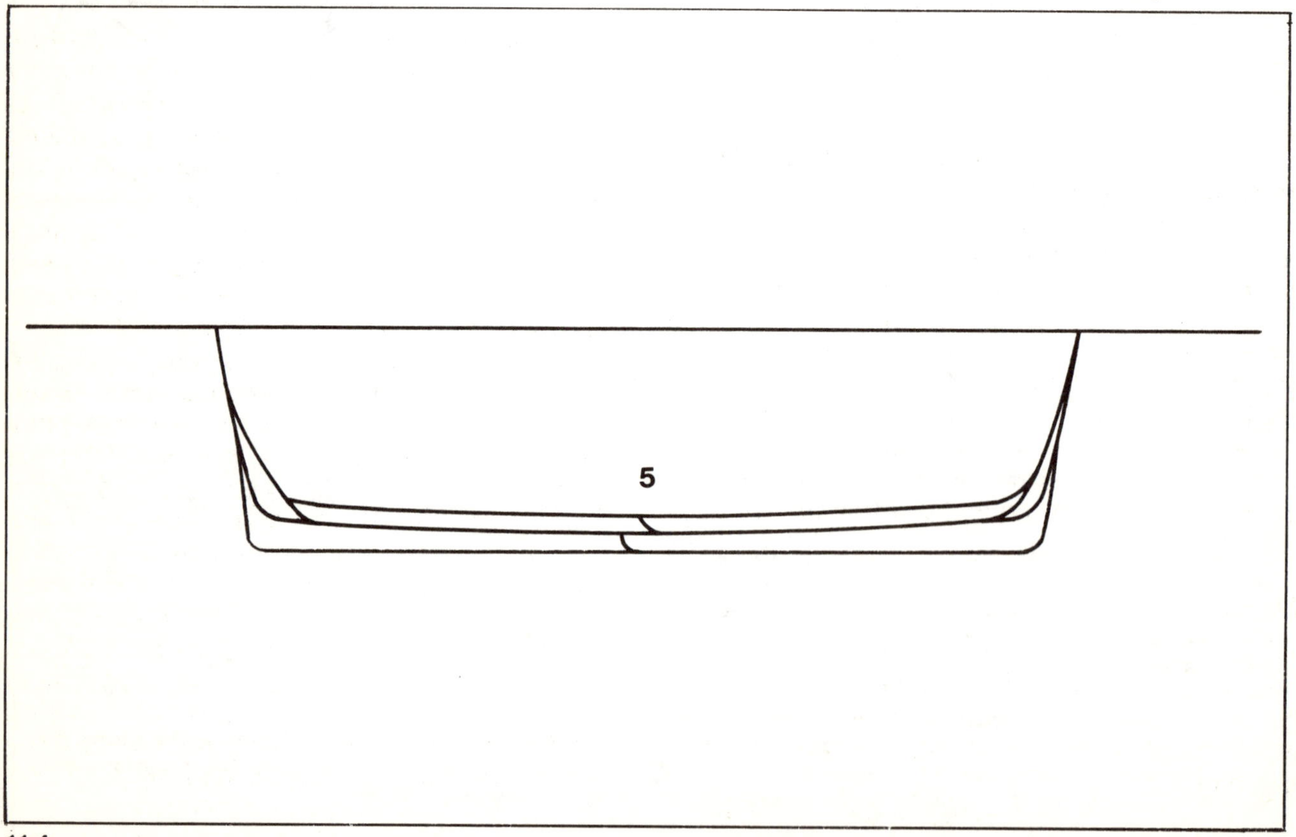

11-4

ing steps 4 and 5—preparing the area and fine grading the bottom—to complete the procedure. These steps are also divided into two stages—the first being with the unit extended, cleaning and fine grading the far section of the trench, and the second stage with the unit retracted, cleaning and fine grading the near section.

Using this procedure will get the best results from the TLB equipped with an extendable dipperstick, and will result in a clean, professional excavation in the shortest time.

Soft Ground Conditions

When working in soft ground, it is easy to fill the bucket quickly while digging. Since there is little resistance against the cutting edges, the bucket can be filled quickly by digging to nearly any depth per cut that the operator desires. If there is little or no chance of utilities being present in the area, digging the soil in deep cuts is acceptable.

The areas where utilities are usually not a problem include open, undeveloped tracts, or on a demolition job where all the utilities are "dead" or out of service. TLB operators don't always have the luxury of knowing positively there are no utilities present, however.

Because of the **threat of utility damage** on the larger percentage of jobs, digging in deep cuts is usually not recommended, regardless of ground conditions. (See chapter on ***Utilities.***) In many other instances, **producing a flat, level bottom is important.**

These situations involve such a large portion of the TLB's **problems,** and the operator's **desired results,** that learning the correct digging technique is absolutely necessary to achieving professional results. In short, digging in soft ground gives the operator two options: he can gouge the soil deeply, filling the bucket quickly. Or, when nearing the desired grade, finish-grading, or in areas where utilities are a concern, he can fill the bucket gradually, in long, grading-type passes.

Hard Ground Conditions

Hard ground conditions sometimes cause problems when excavating. These problems can be compounded by using a large bucket, or worn bucket teeth, or an underpowered machine.

The most effective method of digging in this situation is to make long, scraping passes, taking as much material as possible without **stalling** (or stopping) the bucket. When ground conditions are very hard, the operator may only be able to take an inch or two at a time, scraping from as far as the bucket can be filled with accumulated sluff prior to dumping.

Hard ground conditions also require the operator to understand the geometry of the TLB, and the ways in which the machine produces the most power at the cutting edge. In most digging situations, the bucket should be positioned so that the teeth make contact with the ground at an angle from 30 to 45 degrees (Fig. 11-5A). When hard ground is a factor, this angle should be reduced nearly to the point where the back of the bucket makes contact with the bottom of the excavation (Fig. 11-5B). In this position, the teeth become more effective as they cut through the ground with less resistance.

It should also be understood that the power generated by the TLB in this flat cutting attitude is at a maximum, again due to geometric advantages. In fact, the pulling force produced by the machine when positioned this way is the highest the machine is capable of producing.

Stabilizer Usage and Technique

As we have seen in the chapter on ***Set-up***, learning to use the TLB's stabilizers correctly and effectively is a key ingredient in productive, safe backhoe set-ups. If the machine was limited to digging simple trenches in flat, open fields, there would be little need to discuss stabilizer techniques or their applications. As it is, however, with the exception of the TLB that is used only to dig farm drainage ditches and the like, very few TLB operations are that simple.

One of the most commonly used set-ups is the **tripod set-up,** which involves using the loader bucket and the stabilizers to create a three-point platform from which the TLB may work and maneuver. The two stabilizers are by far the most important elements in providing a safe, stable set-up. To provide this stability, they are designed to extend out from, and behind, the machine itself. This extension feature can often cause a space problem, however. In situations where a trench is close to a wall, for example, the stabilizer (when lowered) prevents the center pivot of the swing from being positioned closely enough to dig a straight trench. If the trench is to be used as a footing, the results must be precise. We will discuss how to avoid these problems, and how the desired results can be obtained in spite of them.

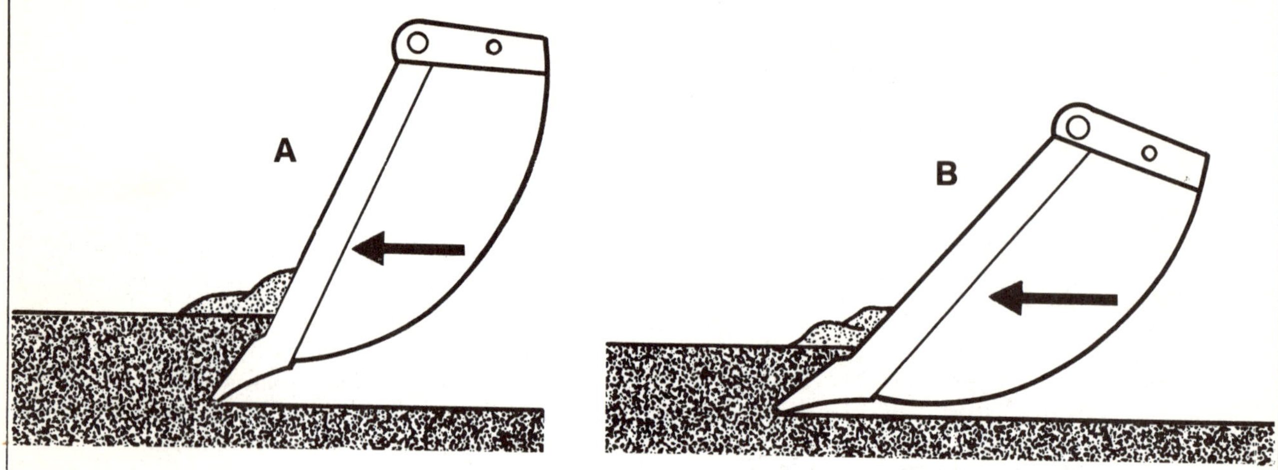

11-5A. This is the desired bucket attitude for standard digging procedures. B) When hard ground is encountered, the angle should be lowered to increase the digging ability of the machine.

Stability. As their name indicates, the prime function of the stabilizers (or "outriggers") is to provide stability when operating the backhoe. This stability is "engineered" into today's modern TLB, and in common working situations, they do an adequate job of stabilizing the machine. There are, however, many situations where machine stability can be suddenly lost. When performing tasks such as craning, material handling, or using the extendable dipperstick, the machine can become overloaded, and will begin to lose balance and overturn (Fig. 11-6). Although there is little chance of the TLB actually rolling over, the operator could be thrown from the machine or tossed violently within the cab, causing injury. Because of this risk, the operator must constantly monitor the weight of the load, and the weight distribution of the machine plus the load.

There are other factors which can affect machine stability during operation. One is the ability of the ground to support the weight of the machine plus the load through the stabilizers. Another is the working width of the stabilizers during operation, which is determined by how far they are lowered. When lowered only to ground level, the width from pad to pad is ten feet (Fig. 11-7). When lowered all the way, the width—and the stability—are reduced (Fig. 11-8). Here we also see that the stabilizers play a part when leveling the machine. When working on a slope or other uneven ground, one stabilizer must be lowered farther than the other, resulting in decreased stability on the lower side. The operator must be aware of these different situations, and how they affect overall machine stability.

Tilting the machine to work around an object. This is a procedure that is used rather infrequently, but can eliminate a problem situation. It is most often needed when the surface cut has been pre-determined, and an obstruction is encountered beneath the surface. As illustrated in Fig. 11-9, one stabilizer is lowered farther than the other and the machine is tilted slightly. Stresses in the tractor and loader can be lessened by lowering the loader so that only one side contacts the ground.

Note: Using this technique will produce an excavation in which one side will have a negative slope. An area such as this should be considered unsafe, and workers should not enter the excavation until it has been properly shored.

11-6

Digging Near a Wall or Fence

Digging against a fixed structure like a wall or fence is a common task for the TLB when working in urban areas. In a few situations, the trench can be moved away from the wall a few feet to simplify the excavation process. However, relocating the trench will be unacceptable in most situations. Footings, for example, must be dug in an exact location, to the exact width and depth.

To achieve results which reflect these exacting standards, the hoe's swing pin must be positioned directly above the center of the ditch. (The closer it is, the more exact the excavation will be.) It is obvious why this could pose a problem for a standard TLB. To ease the excavation of footings and similar procedures adjacent to walls, the **side-shift** or **offset** hoe was developed. This machine is designed in such a way that the swing pin can be shifted to either side of the machine, to aid the operator in lining up the machine over the trench.

These machines, however, are still less versatile than the standard center-mount hoe, and are not commonly found in the United States. In Europe, however, where tight working conditions are the rule rather than the exception, the side-shift hoe is a common sight, and the center-mount is a rarity. Digging next to walls is a fairly straightforward task with the side-shift hoe, so we will instead concentrate on methods that can make these types of excavations easier when attempting them with the standard center-mount hoe.

It should be noted that not all situations are the same, and it won't always be possible to get a proper set-up. If the excavation is for something other than a footing (i.e., less exacting), it can usually be dug without having the swing pin aligned exactly.

In this case, during the digging process, the **swing** may have to be used as the bucket is pulled toward the machine to keep the ditch straight. If the trench width is pre-determined by a pavement cut, or if it must not end up wider than the desired dimension, a narrower bucket can be installed to make this operation easier.

In another example, let's say a trench needs to be dug next to a chain link fence. Sometimes the fence can be loosened and raised to allow the

11-7 When the TLB stabilizers are lowered to the ground, the width of the spread is ten feet.

11-8. When stabilizers are lowered all the way, the spread is reduced and machine stability is lessened.

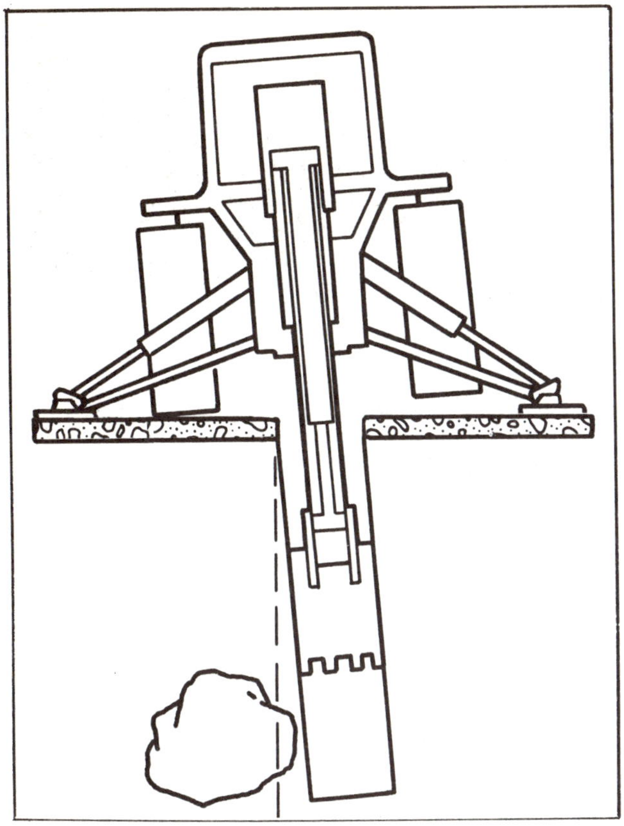

11-9. When an object is encountered underground, tilting the machine will often allow the bucket to work around the object.

stabilizer underneath, greatly easing the set-up and digging processes.

There are some great dangers involved in this type of excavation. When digging near walls, the one thing that should **never** be done is to take away the support or stability of the wall or structure next to the excavation. **Never undermine an existing footing!** Because of the danger of collapse, this is a major consideration whenever digging next to a wall or building.

Set-up techniques. As we noted, the major aspect of setting up for this type of excavation is the placement of the swing pin. Basically, there are three methods that permit correct placement of the machine. The most common set-up problem here is the stabilizer, which sticks out too far to position the machine close enough to the wall.

Method #1. Angle the machine enough so that the swing pin moves closer to the wall (Fig. 11-10). The greater the angle, the closer you can position the pin to the wall (Fig. 11-11). However, as the angle is increased, the arc of the swing is reduced (Fig. 11-12). When this happens, spoil placement can become a problem. Because of the stockpiling limitations of this approach, deep or large excavations will usually be difficult. If no better method is available, the spoil will continuously have to be pulled back to prevent it from falling back into the excavation. This is a tedious process at best. In most circumstances, however, the excavation will be shallow enough so that large quantities of spoil won't be produced, and the swing area will be adequate to complete the job.

Method #2. Another way to place the pin closer to the wall is to reduce the width of the stabilizer spread by lowering them all the way. By combining this maneuver with method #1, the TLB can usually be positioned at less of an angle relative to the wall, increasing the arc of the swing and thus improving the stockpiling range. (Fig. 11-13).

Method #3. The last and least desirable method involves raising the stabilizer next to the wall and angling the machine slightly (Fig. 11-14). In this position, the lowered stabilizer will provide the necessary digging support to the machine. The trade-off is that stability is reduced greatly. Although there is little chance of roll-over, the machine will shake more and overall control will be affected.

Under some circumstances, where there is no question about the strength of the wall, the stabilizer can be extended to align the swing pin correctly **and** stabilize the machine at the same time. (See **digging against a bank.**)

Procedure. When proceeding with the excavation using a combination of methods #1 and #2, the set-up is repeated continuously; achieving this repetitive set-up requires precise placement of the front loader bucket for the purpose of a pivot. This will allow the operator to pivot the machine easily toward the wall, and into the desired set-up.

Once the operator is ready to start, his first step is to determine the angle necessary to allow proper placement of the swing pin. Once the angle has been determined, the machine can be backed into the correct position, **or** the front loader bucket can be lowered in the correct spot, and the pivot method can be used to move the rear of the machine.

There are basically two ways of proceeding with an excavation against a wall. One way is to propel and maneuver the TLB with its hydraulic power. The other way is to set up the machine using the loader drive at each set-up point.

Digging against a bank. This procedure is similar to digging near a wall as far as the set-up is concerned. Because the task is performed

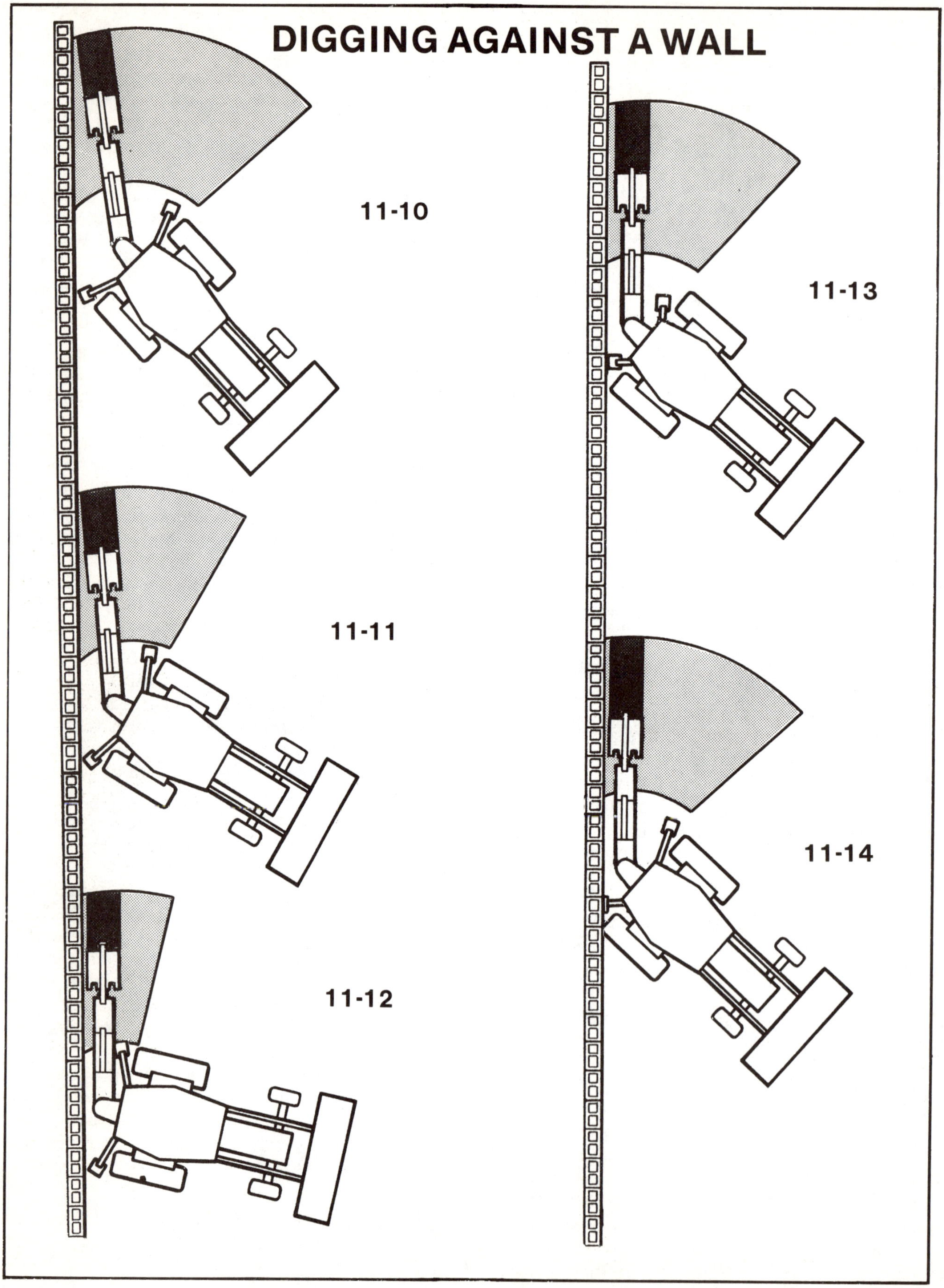
DIGGING AGAINST A WALL
11-10
11-13
11-11
11-14
11-12

with the "inside" stabilizer raised, the machine can be positioned more easily, and closer to the edge. Another advantage here is that the machine may be positioned at a lesser angle relative to the bank, allowing a greater range of spoil placement.

This type of work is most common when excavating hillside foundations. Before attempting such an operation, the slope should be checked carefully for stability. The operator should also check height, angle of the slope, and the safety of the procedure in general.

Specialized Job Tasks

Craning. Although the backhoe is designed mainly as an excavating machine, it has a wide variety of other, more specialized applications, one of which is craning. However, craning is also one of the most dangerous procedures an operator will perform. As has been stated earlier, any procedure new to the operator, or one that is rarely performed, should be executed slowly, and should be well thought-out ahead of time. Craning falls into this category of procedures for most TLB operators. Because it usually involves the help of a laborer for the chain or cable hook-ups, the chance of injury is even greater. These operations **cannot be rushed;** they must proceed at a pace which ensures absolute safety for everyone involved.

Either chains or cables can be used in craning operations, and generally should be rigged as short as possible to prevent the load from swinging out of control. The hook-up should be made in such a way that chain slippage or load shifting during craning are prevented. A few good and bad chain set-ups are illustrated in Figs. 11-15 11-16 and 11-17.

Tunneling

Tunneling with the backhoe may be required in any situation where disturbing or defacing the surface must be kept to a minimum. This could include obstacles such as sidewalks, driveways, curbs and gutters, or street pavement, all of which may have to be preserved undamaged. Or, in some cases, an obstruction such as a pipe may prevent digging straight through as planned. In any event, the dirt must be excavated while working around the object.

Sidewalk areas, for example, are city property in most locations. It is common for gas, water and electrical mainlines to be located under them. As a result, unless utility locations have been posi-

11-15. This craning hook-up is recommended.

11-16. When a shorter chain is desired, several wraps around the teeth will work.

11-17. NEVER hook the chain like this! During the craning procedure the chain can slip around the bucket and the load could drop.

tively identified, tunneling under sidewalks is usually better accomplished by hand, not with the TLB.

Another point to remember is that the operator's sense of "feeling" with the bucket is diminished as the bucket moves in closer to the machine. When tunneling, or any time the bucket is underneath the machine, it is almost impossible to detect a pipe, and avoid breaking it. This is complicated even further because the operator cannot actually see the bucket or what it is doing, because it is underneath him. This takes some getting used to, and requires experience and the ability to control the bucket. Sometimes the operator thinks the bucket is full, and raises the boom only to discover he has over-curled the bucket, and the spoil has nearly all fallen out.

The operator will get better with practice in this procedure. He will be able to "sense" the position of the bucket, and whether it is full or not.

Another approach to this situation is to have someone direct the operator from a position where the bucket is visible. This method is slow, but the amount of material to be removed in most tunneling jobs is relatively small. When the only alternative is hand-digging, the key here is cooperation and patience.

As seen in Fig. 11-18, tunneling is best accomplished by digging under the machine. Caution and common sense must be used in any TLB operation, but especially when undermining the area the machine sits on. Generally, tunneling is safe because the stabilizers support the machine's weight, by extending past the undermined area. Any time a safe set-up is impossible, or when the excavation is eight feet deep or more, however, extra caution is advised, and the operator should consider alternatives to undermining the support area.

Potholing

Potholing is a technique most often used where trenching is required in paved areas, but damage to the surface is to be kept to a minimum. This occurs on a fairly small percentage of jobs, usually in areas where local regulations limit the amount of pavement that can be dug up. Some cities, in fact, charge contractors by the square foot for removed pavement.

Potholing is utilized in one of two ways: 1) it may be a series of holes dug using the same methods described under tunneling, or 2) it may be a series of holes dug at intervals to provide access for boring a tunnel or pushing a pipe with the aid of specialized equipment.

The first method is sometimes used to reduce the need for shoring, usually in sewer line installations of medium-range depth (five to ten feet). In most instances, however, this application is totally unsafe and is **not** recommended in areas where soil is sandy, or contains sand and rocks, or gravel, or any unstable consistency. This technique is more useful in areas where the soil has clay or plastic characteristics. (See Fig. 11-19).

Even in the best soil conditions, however, potholing is not a foolproof procedure. If the area has been excavated previously, the old trench presents a hidden, yet very real danger. With what we learned in previous chapters about the dangers of cave-ins, it is easy to see why care and caution must be used in potholing operations. Shoring or sloping **any** excavation over five feet is still recommended. When excavating potholes, sloping is obviously not possible, which leaves shoring as the only viable alternative.

Using the potholing method allows the operator to dig a continuous trench while only removing potholes five to eight feet long, with the remaining sections between cuts ten to twelve feet in length. Having a TLB equipped with an extendable dipperstick is often useful in this type of excavating.

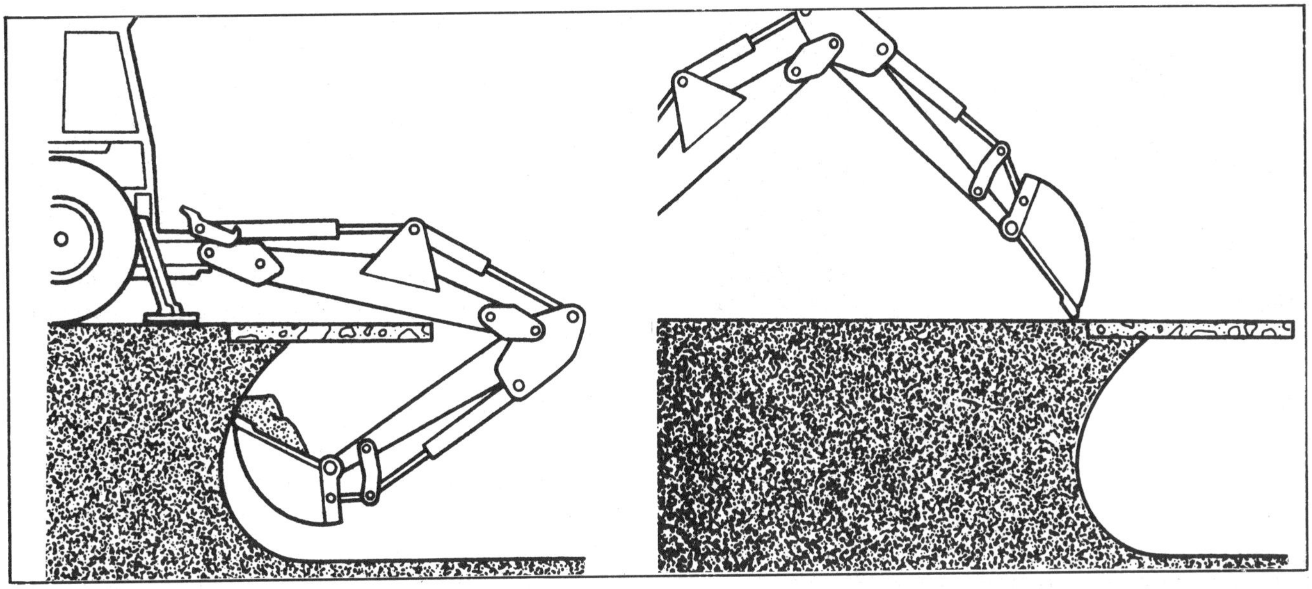

11-18. When tunneling, the backhoe can actually dig under the machine.

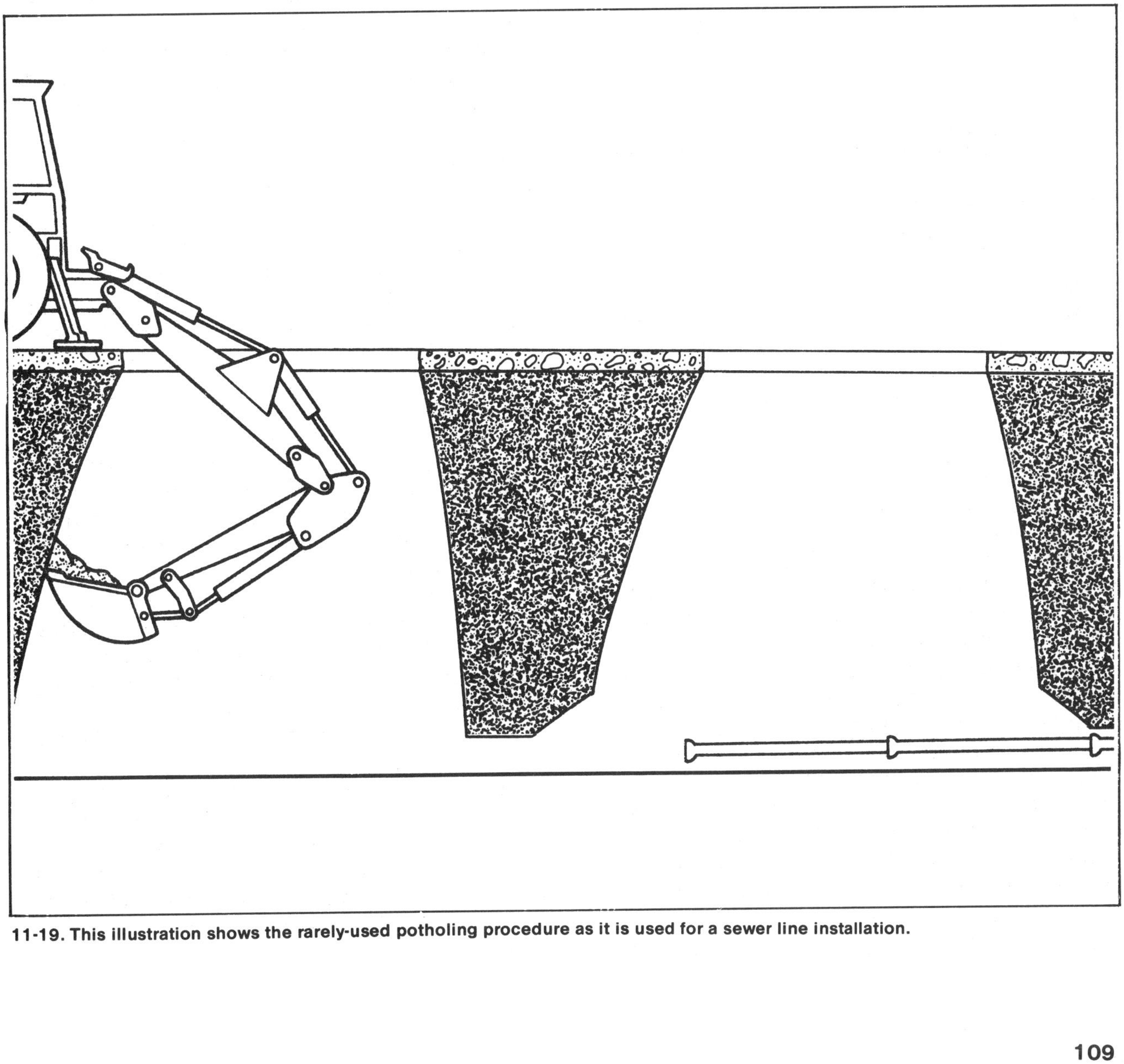

11-19. This illustration shows the rarely-used potholing procedure as it is used for a sewer line installation.

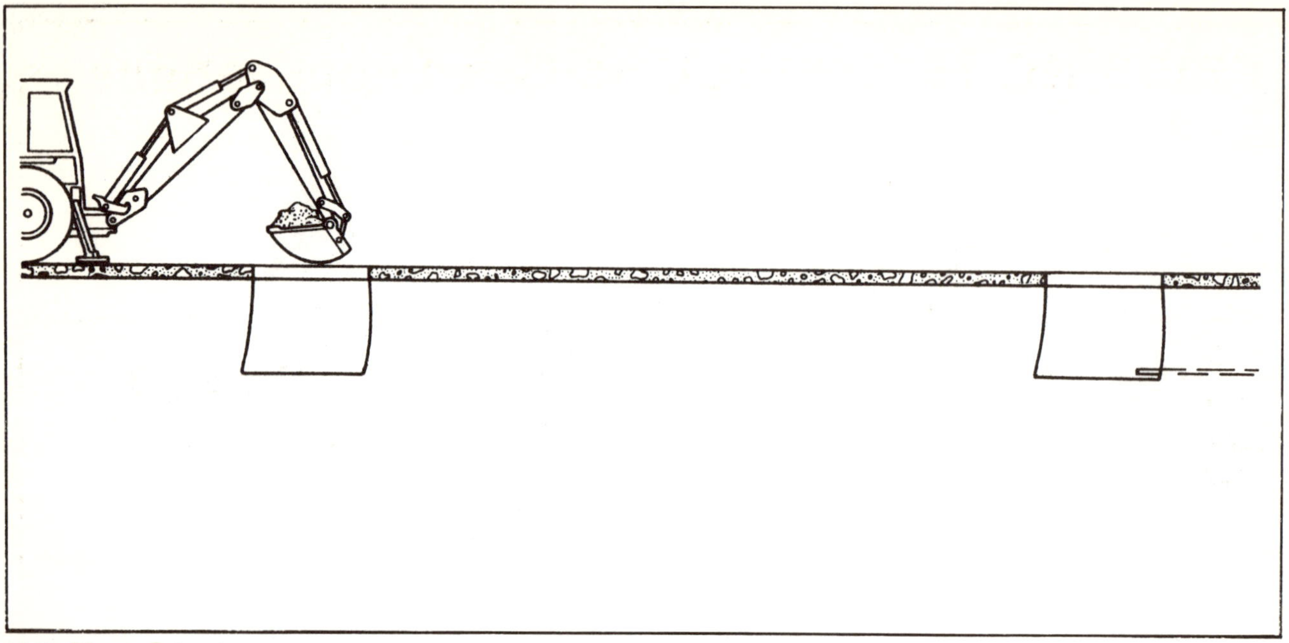

11-20. The use of potholing during a pipe-pushing procedure.

The second potholing application is used in conjunction with pipe-pushing, boring or other specialized equipment. It consists of digging small holes spaced at intervals that depend on the type of equipment being used (Fig. 11-20).

Cribbing

Cribbing is a technique commonly used by TLB operators involved in railroad bed repairs. However, it is also useful in many other situations, such as where digging and cleaning around pipes. As we discuss the railroad applications of cribbing, the method's usefulness in other situations will become more apparent.

Railroad applications: This procedure is used in either of the following two situations: 1) where dirt or mud has filled the spaces between the slag or rocks in the railroad bed, preventing proper drainage and loss of the bed's stability; 2) where spot replacement of ties is necessary. In these applications, the use of a specially-made bucket is necessary to fit between the ties. There are several different styles of buckets, but all are very narrow (usually about eight inches) to fit between the ties.

Cribbing can only be done in areas allowing roadway access along the sides of the track for machine set-up. Throughout the job, care must be taken to position the machine so that it can always back away from the tracks if and when necessary, in as quick and safe a manner as possible. The operator must also be careful not to damage the tracks themselves. Working around railroads is one of the most dangerous TLB operations, and an extra emphasis must be placed on safety at all times, by everyone involved in the operation.

Cribbing procedure: Start by digging down on the near side of the track, from the rail out to the edge of the bed (Fig. 11-21), digging and cleaning both sides of the desired tie at the same time. Then, move over the first rail and continue to pull the material to the rail and then push it under (Fig. 11-22). The key idea here is to start near the machine, excavating a "trench" between the ties that will serve as a sort of transfer-point for material that is being removed, without piling it on top of the ties. Continue working away from the machine, pulling the material under the near rail (Fig. 11-23). Then, as the area up to the far rail is cleaned, the other side is started (Fig. 11-24), and that new material is pushed under the far rail. When the digging has proceeded to the farthest point desired, final cleaning and grading is started at the far point, and proceeds toward the machine until completed.

Cribbing Technique

The technique surrounding cribbing actually involves the motion of the bucket. In this case, the bucket is the focus of all movement, and once it is positioned correctly, the operator simply repeats the bucket motion until all spoil has been pushed under the rail. The most effective

CRIBBING TECHNIQUE: Railroad Applications

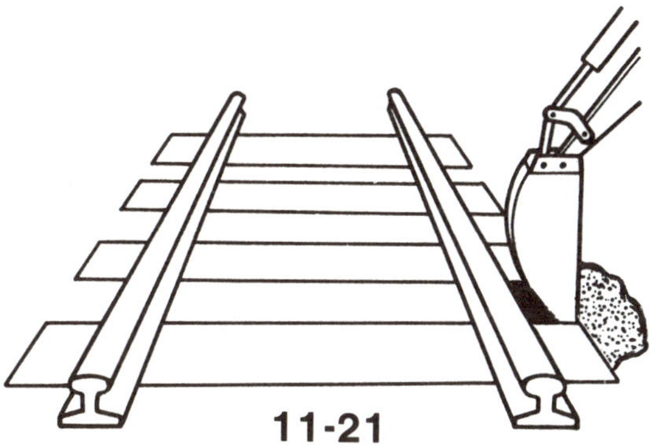

11-21

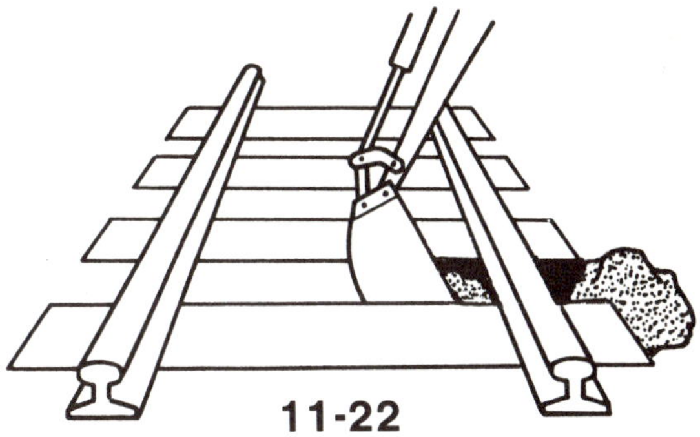

11-22

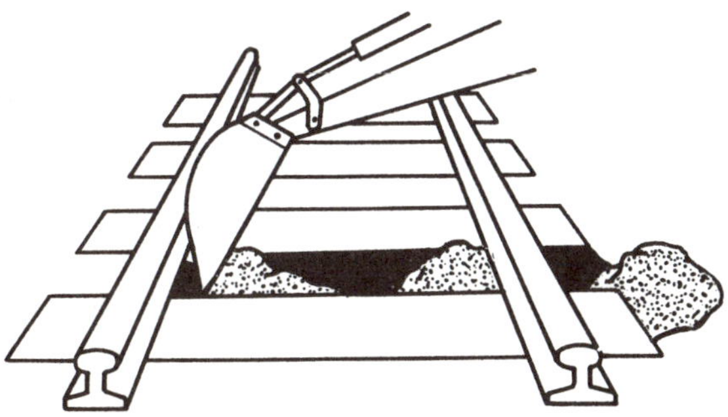

11-23

11-24

approach is to push a small pile of spoil to the rail, and then force it underneath by repeated bucket movements (Fig. 11-25). The effectiveness of this move is enhanced slightly by pulling the bucket to the rail in a straight, even motion. This is most useful when forcing the bulk of the material under the rail.

Then, when making the final grading and cleaning passes, push the material as far under the rail as possible, so that it can be handled effectively and the work continued on the near side of the rail. This technique will obviously require some practice in order to be used effectively.

Other Cribbing Applications

Admittedly, most operators will never use their TLB in railroad applications. Why so much emphasis, then, on a technique that will rarely be needed? Because the technique itself is useful in many other situations which are encountered by all operators. For example, cribbing can be a useful procedure when excavating around utilities. Once a line is located, the cribbing procedure is put into practice. This kind of application, however, requires a very light touch. Utility lines obviously are not as strong as a rail. When the lines or pipes are small, or made of plastic or cast iron, it is wise to clean around them by hand, not with the TLB.

Loading Trucks With the Backhoe

Depending on the hardness of the soil, loading trucks with the backhoe can be quicker and easier than loading with the front end loader. When soil is hard, bucket penetration with the

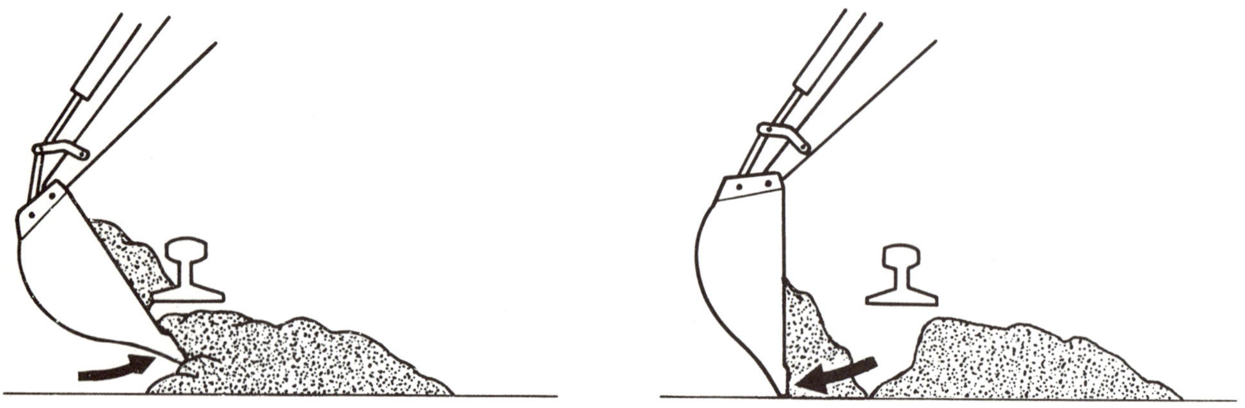

11-25

11-25. A pile of material is pushed to the rail, and by repeating the motion shown, will be effectively "pumped" under.

loader may be difficult. In this situation, the soil can be loosened with the backhoe (Fig. 11-26). Or, if a safe set-up is available, the truck can be loaded "pit" style, by positioning the hoe above the truck (Fig. 11-27). When using this technique, the depth of the cut should range from six to eight feet. A cut of this depth will provide a safe set-up, and provide enough material so that time will not be wasted due to frequently moving the machine forward.

When soil is soft, such as loose sand or gravel, or when the earth is moist and muddy, maneuvering the loader may be very difficult. When the loader bucket is full, the weight of the load has two effects: 1) The weight will force the front tires into the ground, and 2) the amount of weight on the rear tires will be reduced. This, in turn, will cause the rear wheels to lose traction. This is another situation where loading with the hoe may be more efficient, as well as safer.

To ensure efficiency, the machine should be fitted with the largest size bucket available, up to three feet. Also, the bucket adjustment should be set to allow for maximum curling. Using a larger bucket, it should be pointed out, can cause a stability problem. If this happens, either position the truck closer to the backhoe, or use a smaller bucket.

Loading trucks while digging an excavation. This is a fairly common procedure. When the operator knows that spoil is to be trucked out of the area, he should try to load it as it is removed from the excavation. Proceeding with the job in this manner will greatly ease clean-up, and the job will be completed more quickly. When the backhoe is used for trenching, and will be moved forward several times before filling the truck, Fig. 11-28 illustrates a useful set-up for the truck and the backhoe. In this set-up, the truck can move forward easily with the hoe, and the two vehicles work as a team.

If the hoe is digging a large excavation, and can fill the truck from a single set-up without moving forward, Fig 11-29 illustrates a set-up that will help the operator distribute the load evenly in the bed of the truck. One prime consideration, of course, is to prevent damage to the truck (see Figs. 11-30 and 11-31).

Material Handling

Sometimes, while on the job site, the TLB operator will be asked to move, load or maneuver an object or objects. Any large object too heavy or awkward to be moved by hand could be included. When such a request is made, the operator should study the entire scope of the task, from beginning to end. The **operator** is responsible for the safety of himself and others, to the machine, and to the work area. And he alone knows what his machine is capable of. Here again is an example of a situation new or unusual to the operator, and thus the dangers are increased.

Most of the objects which the TLB operator is asked to move will be relatively simple to handle, while others will be difficult and dangerous. Some objects should not even be attempted. It is up to the operator to determine the degree of

11-26. When soil is too hard for the loader, the hoe can usually loosen the material adequately.

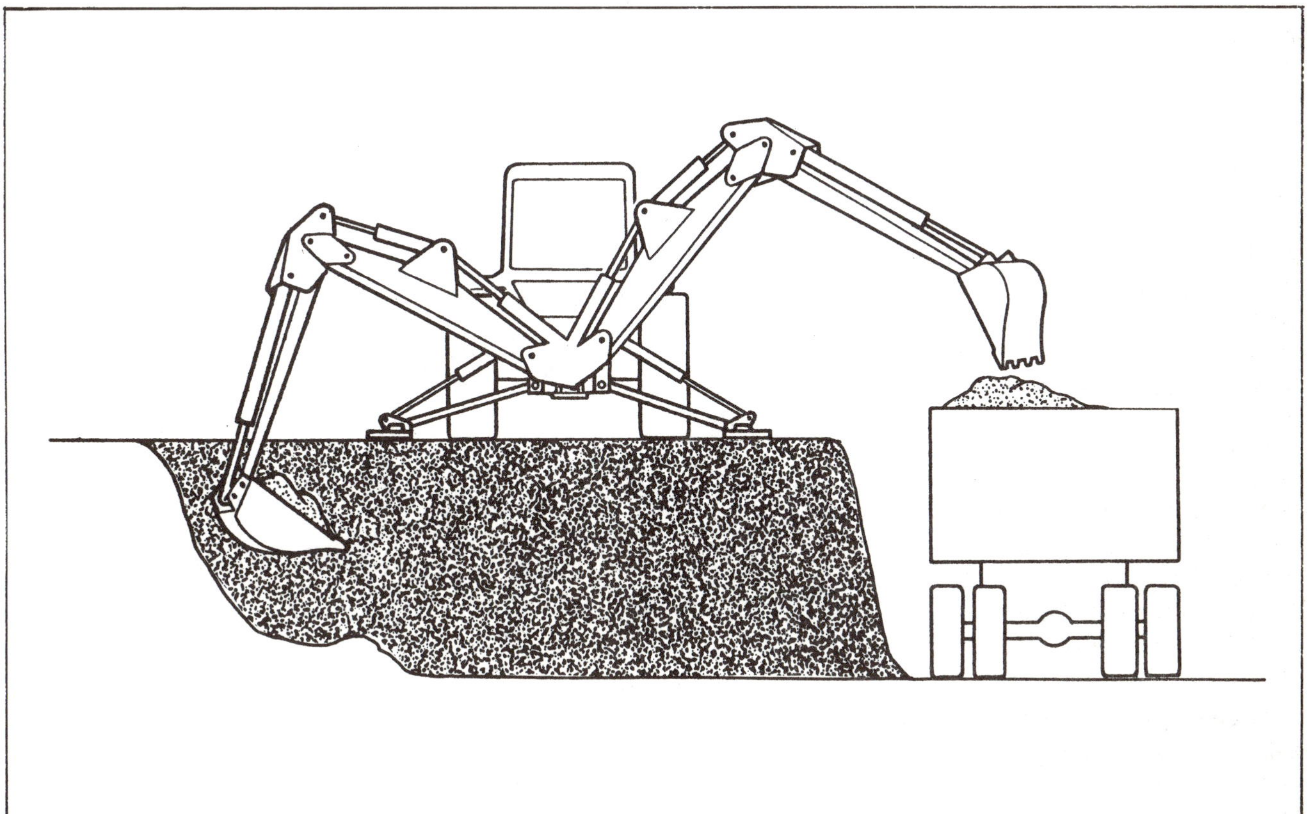

11-27. Loading trucks with the hoe can also be effective for "pit style" loading.

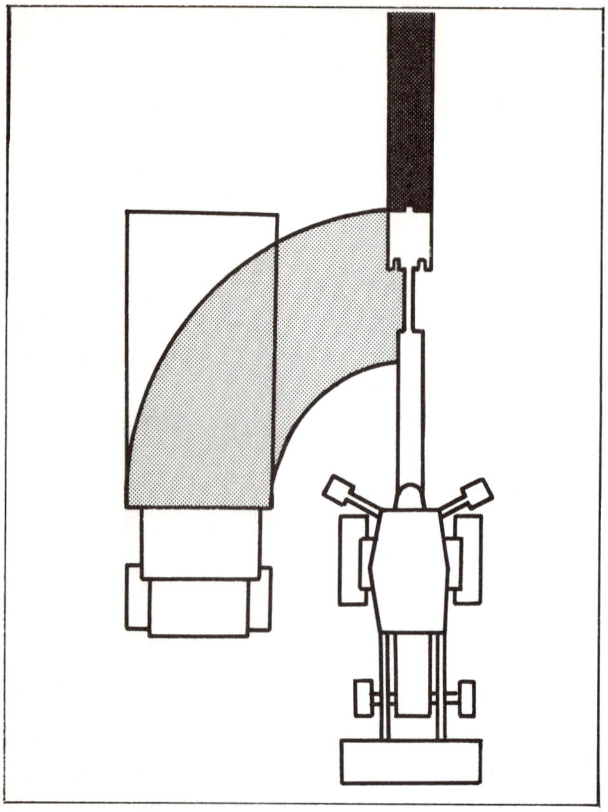

11-28. By using this truck setup, the truck can easily move forward with the machine as the work progresses.

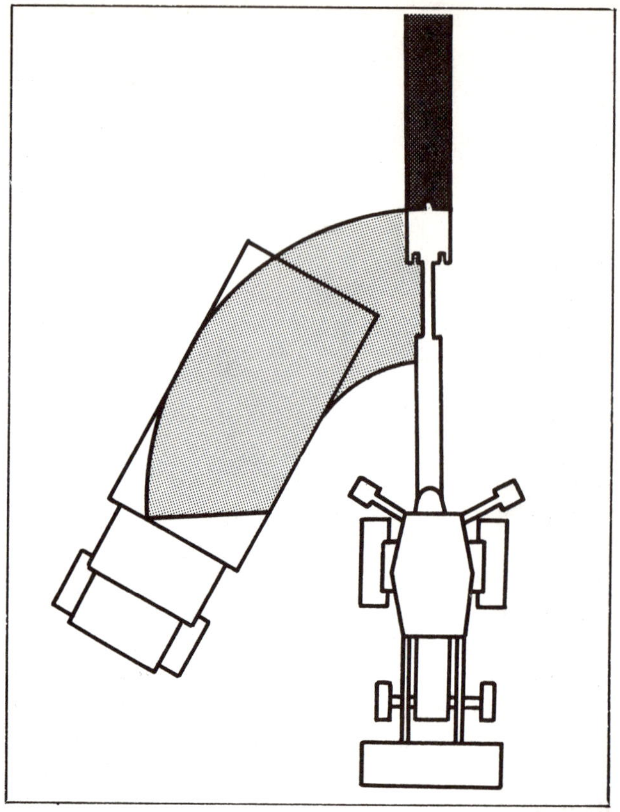

11-29. Using this truck setup will ease the distribution of the material in the truck.

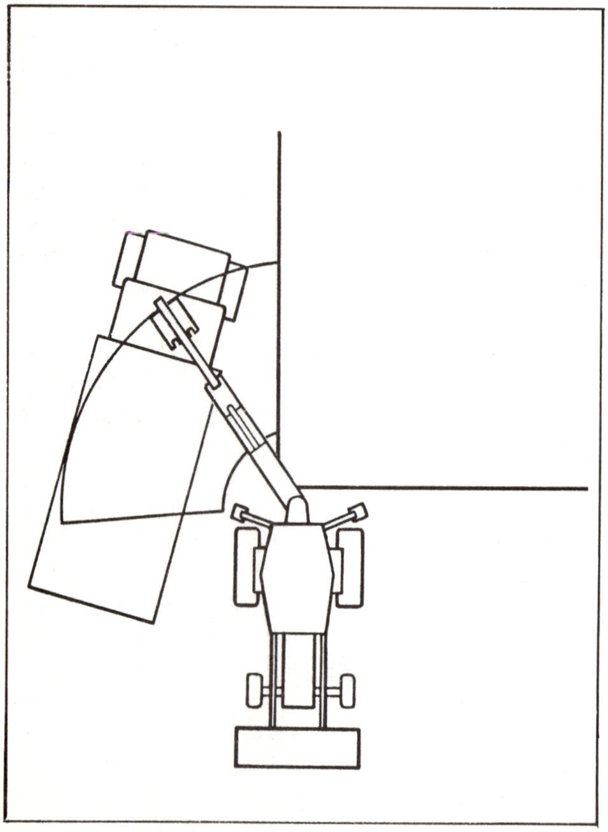

11-30. Never swing a load over the cab of a truck.

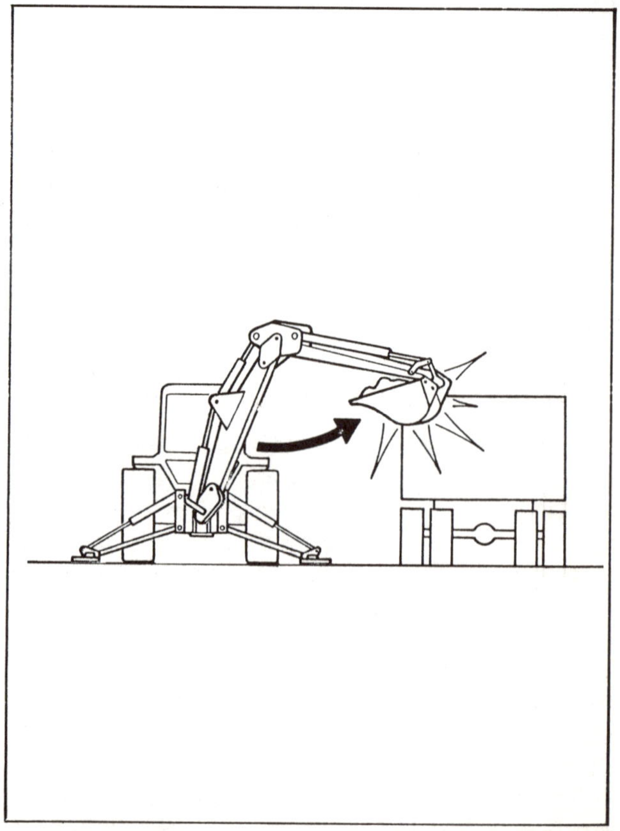

11-31. When lifting and swinging the load, be careful to clear the corner of the truck.

difficulty in each material handling job.

The main element of avoiding problems, once again, is **control.** Controlling an object means moving it exactly as the operator desires. This is best accomplished by maneuvering the bucket directly under the object's center of gravity and pinching it tightly between the bucket and crowd. The object will then be securely held and **controlled** by the operator.

When handling odd-shaped concrete chunks or large rocks, moving these objects **smoothly** is very important to prevent shifting of the load, and losing the machine's "grip" on it. This is especially true when handling or loading out concrete. As the boom is raised, the bucket naturally slants down and the piece must be gripped firmly to prevent it from falling out.

Handling these types of materials requires a very delicate touch on the controls to **slowly** and **smoothly** move the object. Whenever the operator tries to pick up any object, whether a piece of concrete, a rock, or even a telephone pole, he must consider its center of gravity. Think about how the object's weight is distributed into the ground where it rests. Some tricky maneuvering is often required to get the bucket under the object and "grab" it in a way that keeps it from falling out. This type of maneuvering often separates the men from the boys, so to speak, because it requires total bucket control. It's not the same as digging a trench or hole, with the bucket going through normal digging cycles. Digging, scooping, swinging, and dumping are rhythmic operations and smoothness is easy to maintain. When picking up objects, however, there is no rhythm, and maneuvering the bucket and the object often proves jerky and awkward for the inexperienced operator.

Landclearing Procedures

Although the TLB is not the best choice for this type of work, it can be used effectively for small jobs involving the clearing of trees or brush. (Track loaders or bulldozers are better suited for this type of work.) This may include the removal of small shrubs, bushes and other types of lowlying vegetation, small trees and saplings, or larger trees up to a foot in diameter. Removing trees larger than this shouldn't be attempted, because they are simply too much for a machine of this size to handle.

The entire procedure should be carried out so that the downed trees and brush won't be in the machine's way as clearing proceeds. Trees should be downed in an orderly fashion, and piled neatly so removal will be easier when the time comes. Fig. 11-37 illustrates how clearing should progress, and the proper positioning of downed trees. After the first swath has been cut through the area, the procedure will become easier on succeeding passes. The first pass should be twenty to twenty-five feet wide, piling the downed trees on both sides of the cut. Subsequent passes should be slightly narrower (fifteen to twenty feet), and material should all be downed on the side that has been cleared (Fig. 11-38).

Clearing Techniques

Brush and bushes. This is a straightforward procedure, using the clearing patterns described above. The teeth of the bucket, and the "tooth grab" method can both be used effectively. The bucket teeth can be used to "comb" or "rake" the area clean of this type of vegetation. Also, the tooth grab is very useful when pulling out bushes without digging up the area and creating unnecessary holes. When clearing this type of material, it is easy to position the spoil on the side of the swath, which simplifies succeeding passes through the area.

Saplings. Young trees up to three inches in diameter can be removed by using the same procedures outlined for brush and bushes. Again, the tooth grab method can be useful, as long as the trees aren't brittle. If this is the case, and the trees won't stand being bent to the extent necessary for using the tooth grab method, they can be removed using the **push-pull method** described below. If they are flexible, however, they can be easily removed and stockpiled in one motion, using the tooth grab method.

Larger trees. When trees four inches in diameter and larger are to be removed, the most useful method is the **push-pull method.** The first step is to push the tree over and away from the machine (Fig. 11-39), and then loosen and free the remaining roots from the ground by pulling at the stump (Fig. 11-40). When using this method, the machine's pushing ability can be increased by rolling the loader bucket down all the way, and lowering it firmly to the ground. This will stabilize the machine and prevent it from being pushed instead of the tree. This method is preferable to using the machine's swing to down trees, because the trees can be controlled more effectively, and the push-pull approach is generally safer than using the swing.

The direction in which the trees fall can often

11-32

11-33

HANDLING LONG MATERIAL

11-32. When handling long material, the pick-up point must be at the center of gravity of the objects. Here the stabilizer serves as an effective back-up.

11-33. Here the stabilizer is raised to force the ties onto the bucket. The weight of the machine is supported by the bucket.

11-34. Balanced on the bucket.

11-35. The load is positioned close to the machine for stability.

11-36. The stabilizer remains down to ensure stability. The machine travels very slowly on flat, level surfaces *only*.

11-34

11-35

11-36

be a problem, and the operator must be able to control the direction of the fall, or the procedure can be unsafe. This means the operator must be able to make the trees fall exactly as he desires. If a tree falls in any other way, the operator is not in control and the job can become unsafe. There are several factors that determine where the tree will most likely fall. The operator must take these into consideration and judge the best direction in which to attempt the fall.

If the tree is **leaning** in one direction, it is very difficult to make it fall in another direction. In this case it is best to fell the tree in the direction of its natural lean. If this isn't possible for some reason, other means may have to be found to bring it down.

Also, if there is any **wind** on that particular day, the operator must take it into consideration. The strength and direction of the wind can change the plans of the job completely. If possible, the operator can use the wind to his advantage, by planning to down all trees in the same direction, with the wind.

Larger trees (eight inches or larger at the trunk) may be difficult to push over, and may require some preparation, such as cutting some of the roots first. However, it is important to cut the roots on the correct side (Fig. 11-41). If the operator wants to push the tree over directly away from the machine, he should cut the roots on the near side of the tree. By doing this, there will be little resistance when he pushes the tree over. Remember: cut the roots on the **opposite** side of the tree from the desired direction of the fall.

Stump Removal

Another common land-clearing task is the removal of tree stumps. Depending on the type of tree, stumps up to six or eight inches in diameter can be removed with the hoe without having to dig around them. Some types of trees, however, have stronger root systems and will require digging to loosen the stump. If possible, trees of this size and type should be cut higher on the trunk so that the TLB will have more leverage when attempting stump removal. If the tree is cut off five or six feet from the ground, the remaining stump will provide leverage in the removal process. If the stump cannot be removed using the push-pull method, then digging around it and severing the roots will be necessary.

Digging out stumps. The object of this procedure is to loosen the stump by removing the soil surrounding it, and to cut the roots that hold the stump in place. The rule for this procedure is

11-37. When clearing trees, the first pass through the area should be as wide as possible, pushing trees to both sides of the path.

11-38. On later passes, about half the original width will work well. This will allow the operator to push the trees downed earlier away from ones still standing.

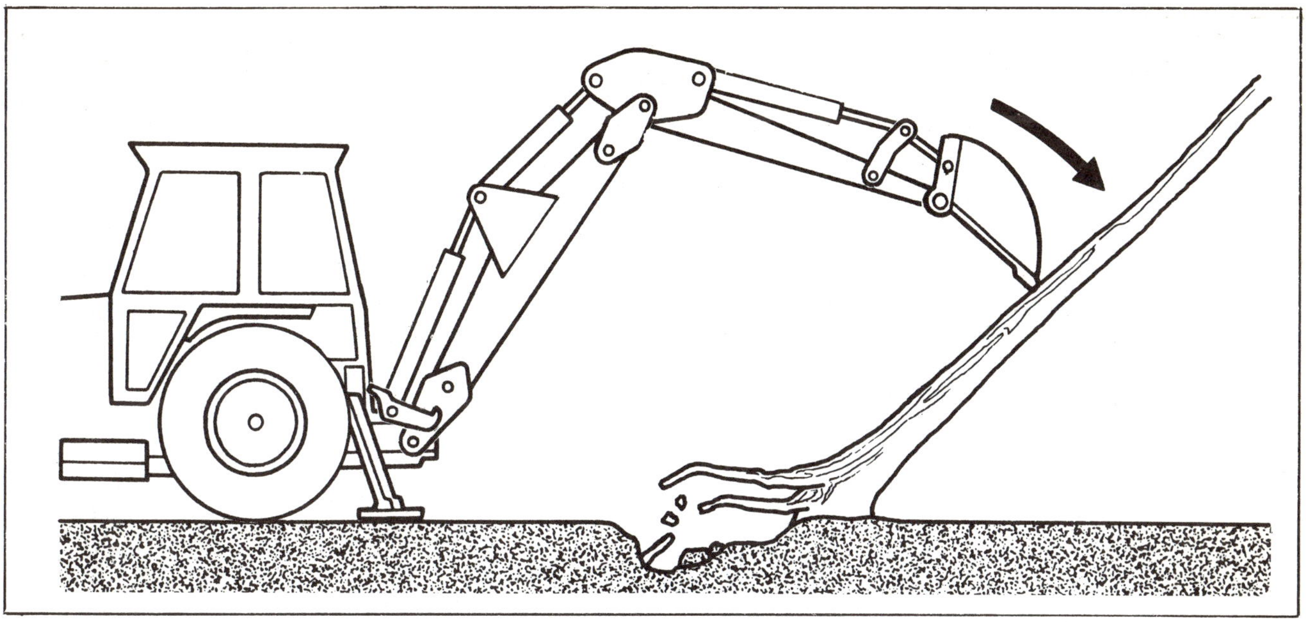

11-39. PUSH: The tree is pushed away from the TLB and down to the ground.

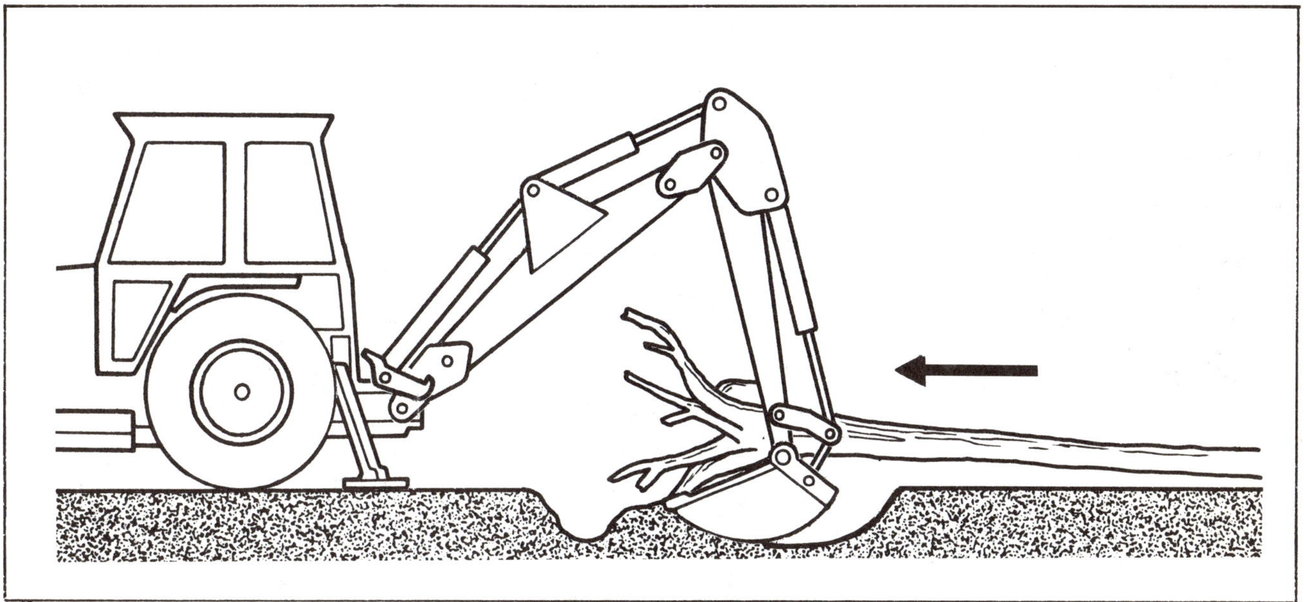

11-40. PULL: The bucket is used to break the roots and pull the tree toward and around the machine to the side.

that the required size of the hole will be about four times the diameter of the stump. So, a one-foot stump will require a four-foot hole in order to loosen it. A stump with a diameter of two feet will usually require a hole measuring eight feet across, and so on. It is easy to see that as the size of the stump increases, the project gets much bigger very quickly.

A stump with a diameter of three feet will require a hole about twelve feet across to get to where the roots can be cut or broken with the backhoe. However, there will often be problems when trying to remove stumps of this size. Even after loosening the stump, the backhoe often doesn't have enough power to lift the stump from the hole.

If this is the case, there are a couple of options open to the operator. One option is to dig an even larger hole next to the stump and simply bury it there. This is probably the easiest alternative, but it is often unacceptable.

Another way to deal with this situation is to fill in under the stump and continue to roll it back and forth. By replacing more dirt back into the hole, and working the stump back and forth, it

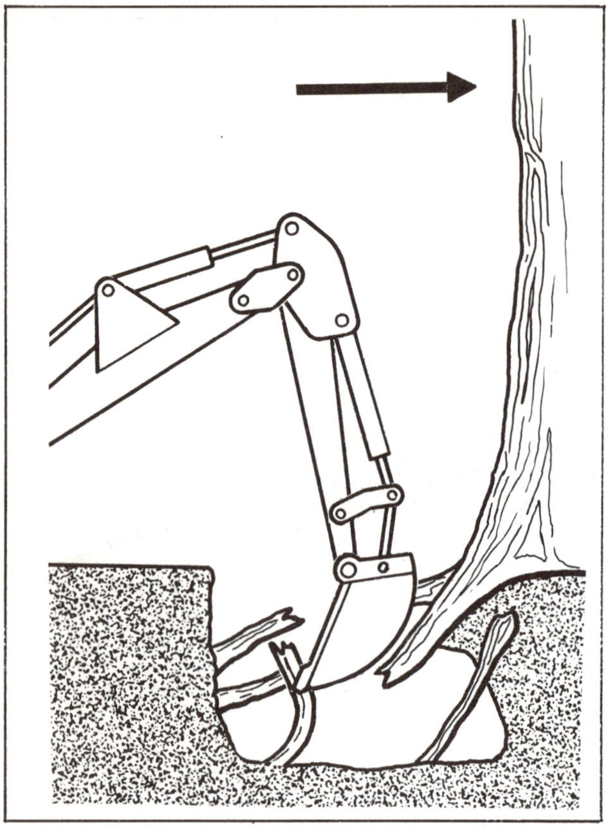

11-41. Cutting the roots on the near side of the tree will allow it to fall away from the machine.

can be slowly raised until it can be pulled out of the hole. This method is also useful when attempting to remove large boulders from a hole.

How to Dig an Arc

Although digging in an arc is a task not often encountered, it is one the operator should be familiar with and capable of performing. It is most often called for when curved walls are to be built, or when a pipeline requires a trench to change direction gradually. When the excavation is to be used for a footing (for a wall), the same rules apply as for any other type of footing. The main consideration is the layout, which must be carefully laid down to the exact specifications, and marked on both sides of the footing. As digging begins, the operator should keep in mind the fact that the smaller the radius of the arc, the shorter the section is that can be dug from each set-up.

As shown in Fig. 11-42, by varying the length of the pass in each set-up, the arc can easily be adjusted to the desired radius. After the excavation has been "roughed out," it will **almost** be down to the desired depth, and will have "corners" and "flats" in the arc. The next step is to make several floating passes with the bucket, using the swing to carve off the corners and

Text continues on page 124

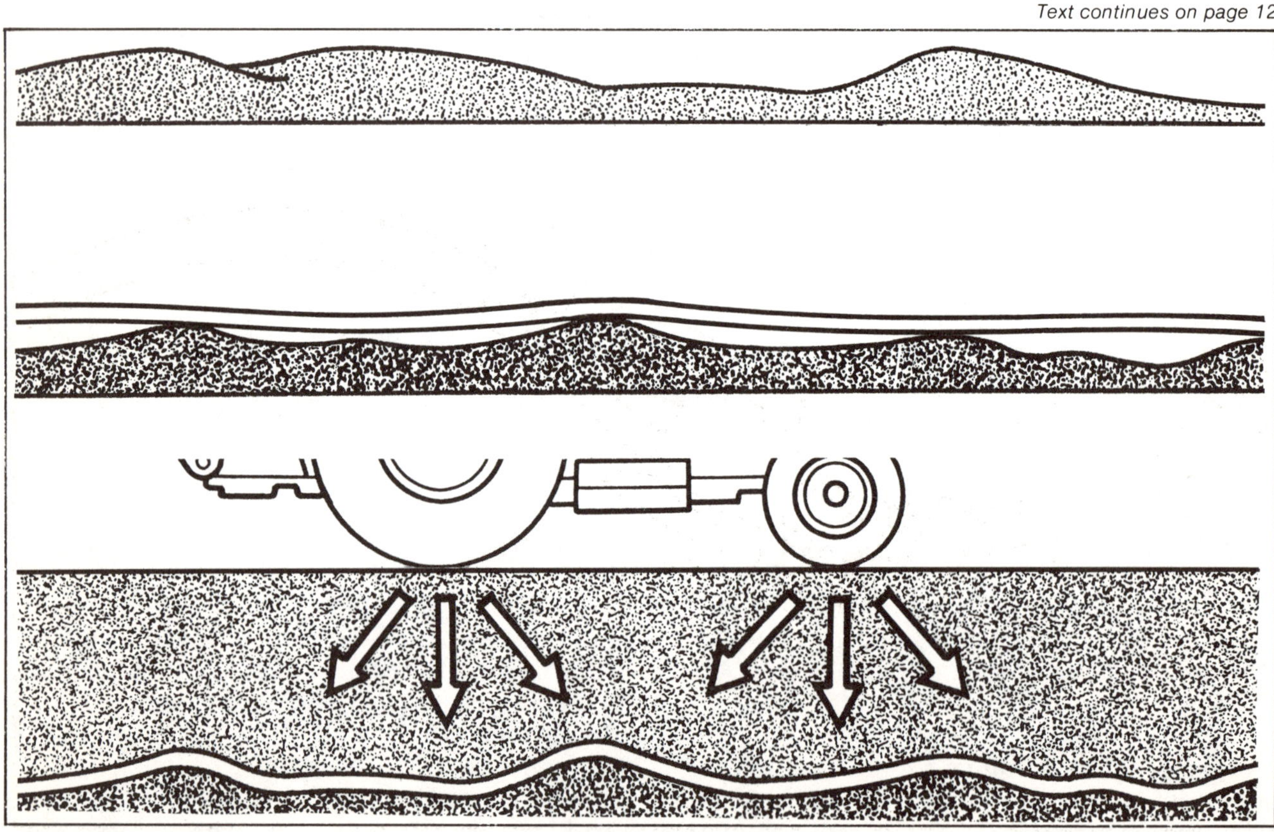

11-43. Conduit laid in a poorly-graded trench will become kinked during compaction.

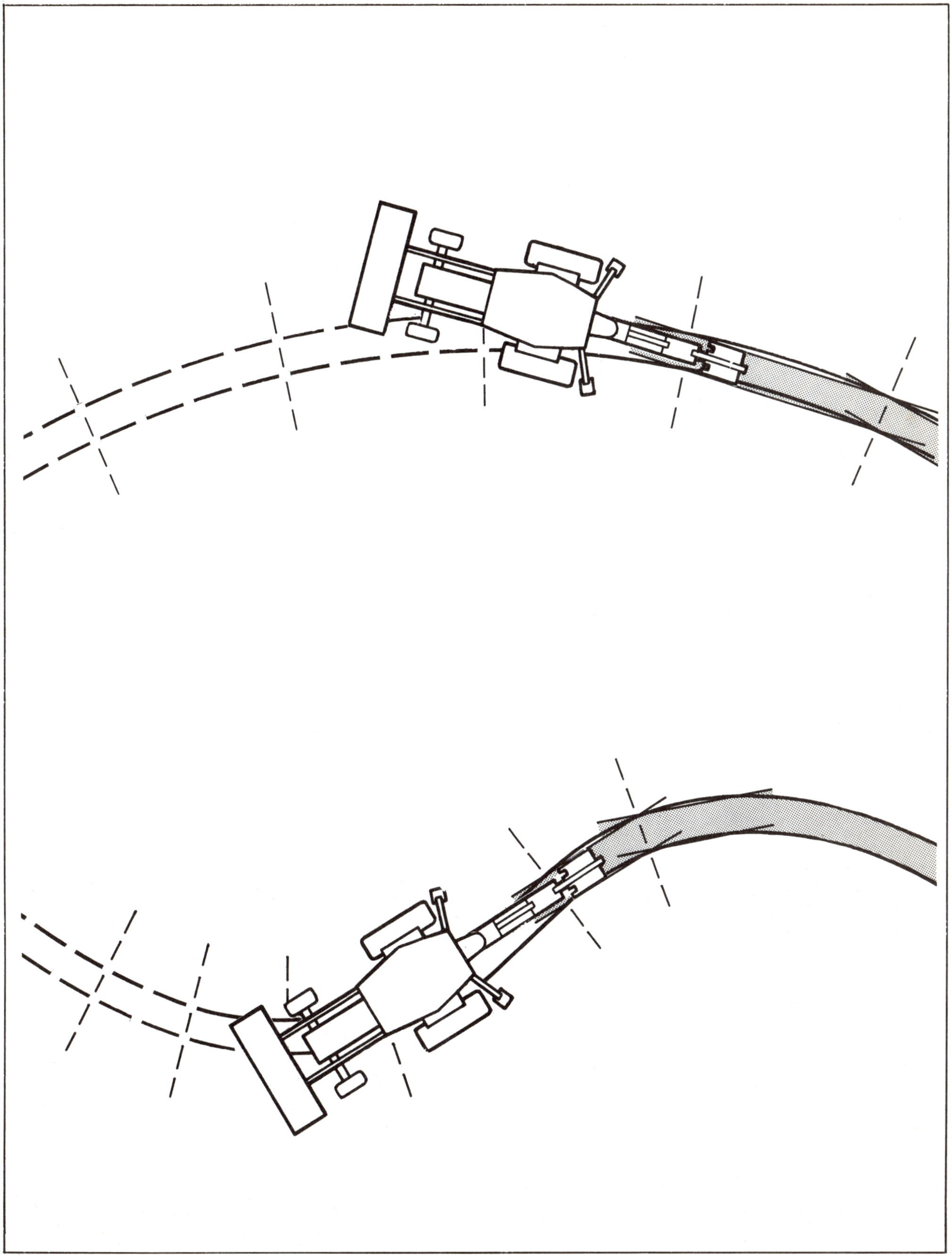

11-42. An arc is actually dug in short straight sections. The smaller the radius of the arc, the shorter each of these straight sections will be. After the section is "roughed out," the sides are smoothed and the bottom is graded.

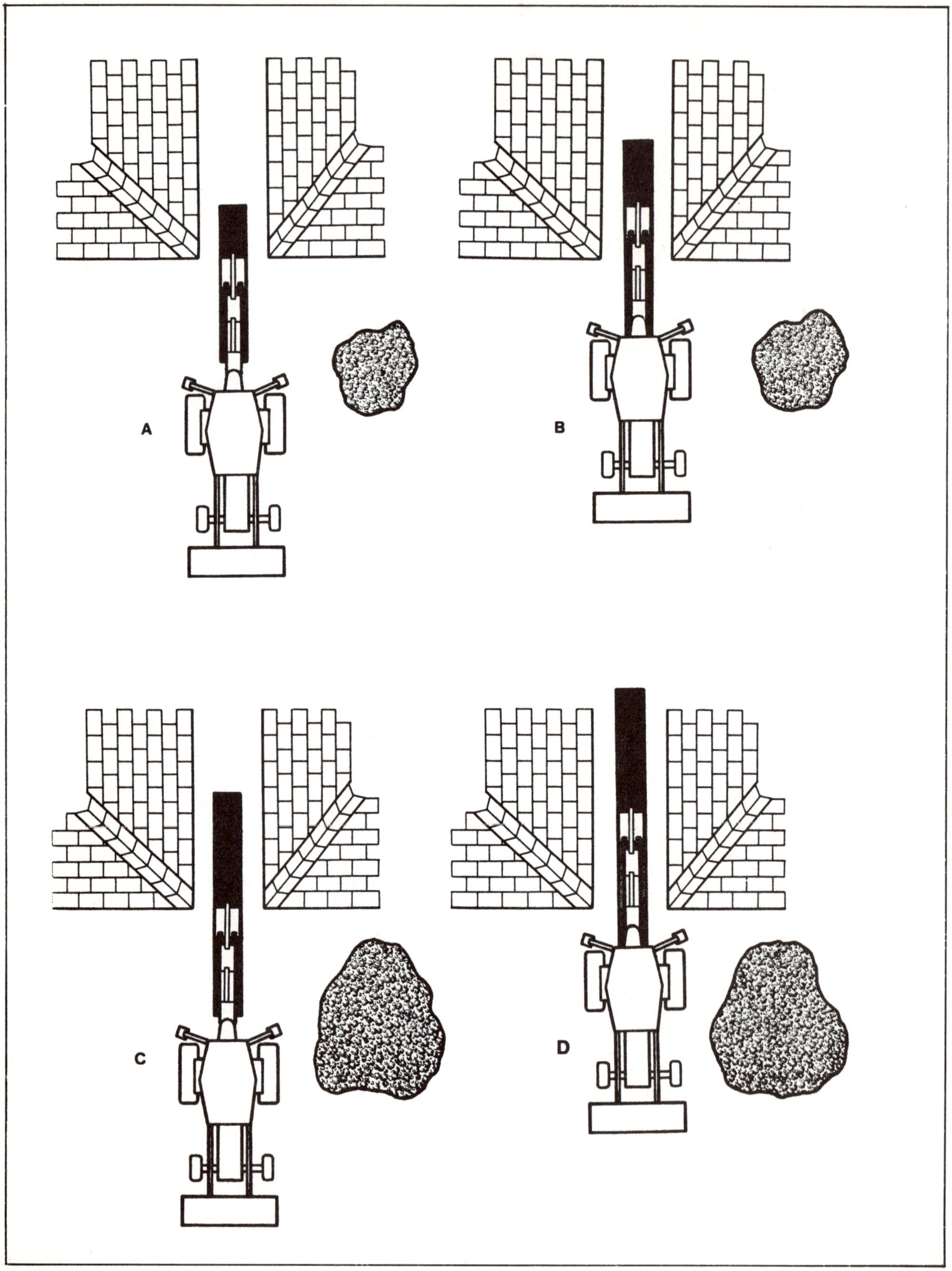

11-44

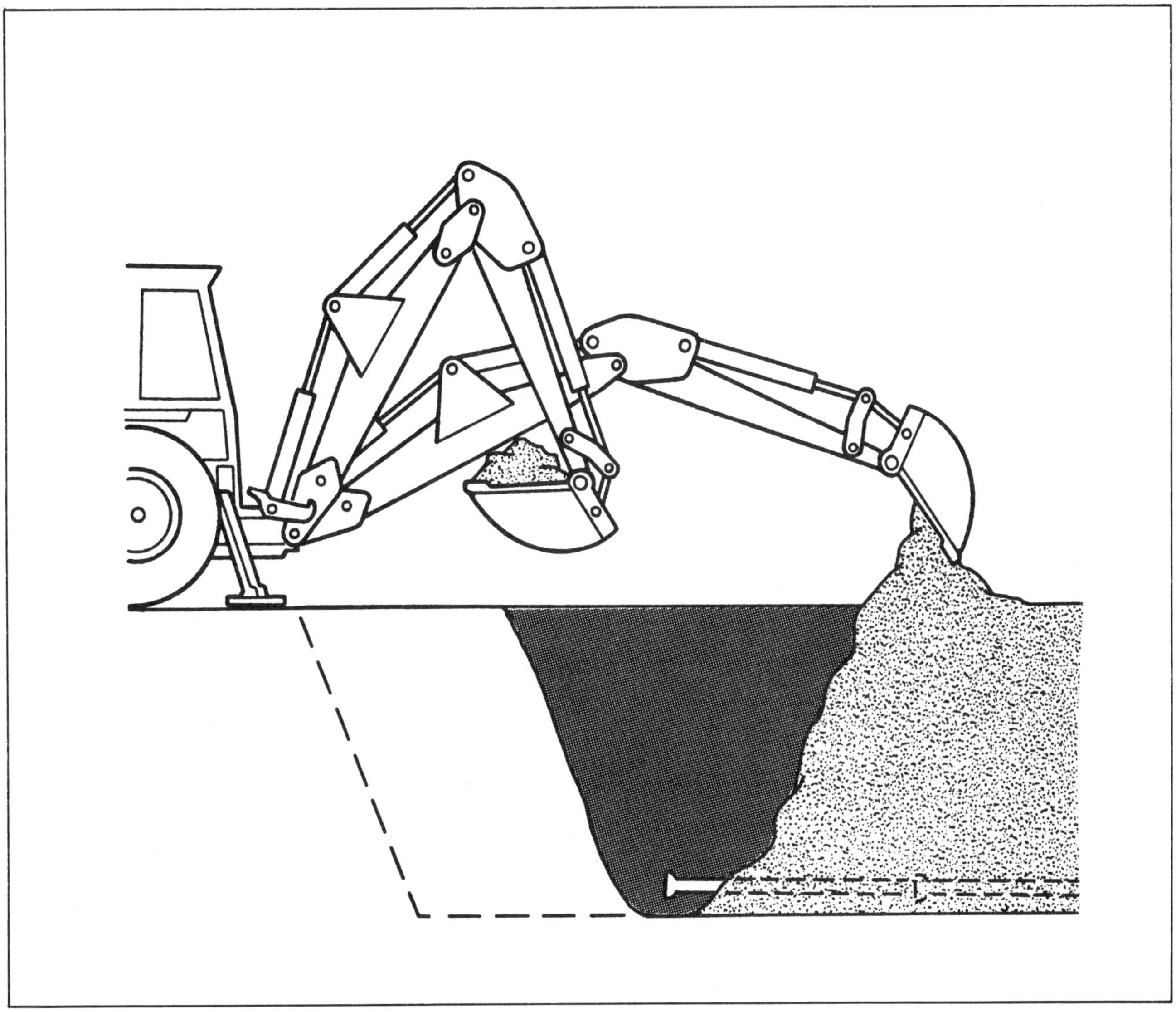

11-45. When digging a trench in stages, the new spoil is placed over the section of pipe just laid.

Fig. 11-44

USING THE TRENCH FOR SPOIL TRANSPORT: As seen in Fig. 11-44, the desired trench is to be dug between two buildings which are too close to accommodate the TLB in between. The machine is set up far enough from the buildings to allow for the swinging and stockpiling of the spoil (A). The area dug from this position will serve as the area where all of the spoil from the entire trench will be placed. As the operator moves the machine back, the newly-dug spoil will be pushed under the machine (B). When the available space under the machine is full, the machine is moved out to the first position. Again, the operator removes the spoil from the trench and places it in the new stockpile area. Then the machine is moved back as far as possible and the trench is dug as far as the backhoe can reach (D). Again the spoil is pulled out into the area where the material can be stockpiled.

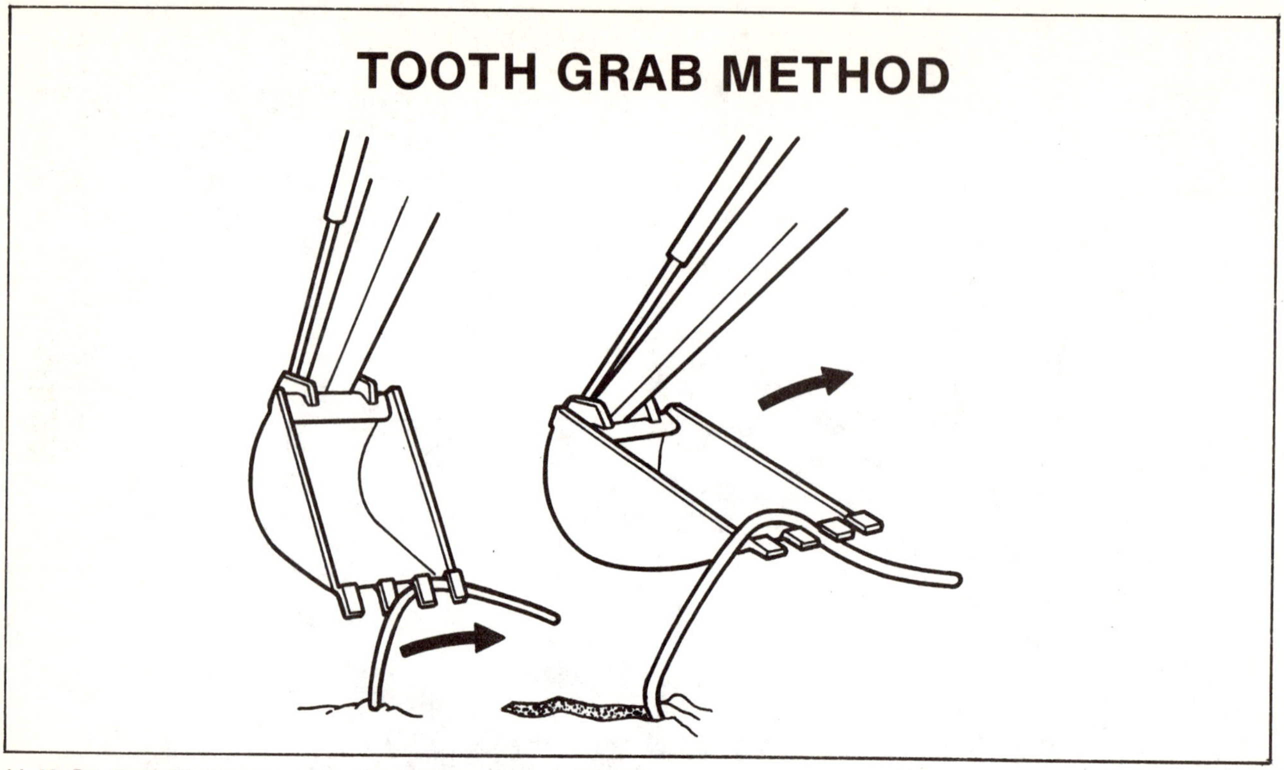

11-46. Starting with the object pointing straight up, the teeth are wrapped around, resulting in a grip that usually won't allow any slippage.

slightly round out the flats (or straight areas). Then, after the excavation has been rounded off, the remaining spoil is scraped out and the footing is cleaned to the desired grade.

Another example of a task that will require the digging of an arc, as we noted, is when excavating for a pipeline. This is especially common when electricians are to lay plastic electrical conduit. Here, a very smooth bottom and a gently curved arc will be required. This will eliminate any dips or rises that could cause trouble later, when the wires are pulled through the conduit. Such irregularities could kink the pipe later, after back-filling and compacting the area, so that the wires will not pull through (Fig. 11-43).

What to do When No Spoil Placement Area is Available

Although spoil placement can often be a problem, there are times when the trouble can be extreme—there will be no place to put it all, or the work area will be so narrow that the swing cannot be used. These situations are rather infrequent, which is good news since dealing with them can be difficult and time-consuming. There are, however, two methods which can be used to ease job completion in these situations. The first method, using the trench for spoil transport, is illustrated in Fig. 11-44, while the second, digging the trench in stages, is explained in Fig. 11-45.

Tooth Grab Method

This is a very useful procedure which can be used in a variety of situations, as we have noted. It can best be used to grab reinforcing steel, small diameter posts and pipes up to two inches in diameter, as well as shrubs, bushes and small trees. Fig. 11-46 shows the exact motion used to "grab" the pipe, tree or other object, while Figs. 11-47 through 11-49 demonstrate the strength and effectiveness of this method.

Note: These photos are meant as a demonstration of the effectiveness of this method only; we do not recommend handling concrete in this manner. For more on concrete handling and the tooth grab method, see chapter on ***Demolition.***

Connecting Trenches

This is a common TLB procedure, and involves starting a trench from both ends and digging toward a meeting point at the center (Fig. 11-56). The need for this approach occurs most often when a trench is being dug between two buildings, or when digging footings for a large structure.

11-47. Starting with the rod between the teeth, the bucket makes a circular motion . . .

11-48. . . . and the rod is gripped securely between the teeth.

11-49. This is intended only as a demonstration of this method's effectiveness. Handling concrete in this manner is not recommended.

11-50. This procedure is recommended for trenches two feet deep or less. When approaching the trench, the bucket is lowered and the weight of the front of the TLB is supported by the bucket.

11-51. Once the weight is supported by the bucket, the machine can be moved forward until the rear wheels approach the trench.

11-52 The stabilizers are then lowered and will support the rear of the machine until it is moved forward.

11-53. The machine is then pushed forward with the backhoe until . . .

11-54. . . . the weight is supported by the rear wheels and the stabilizers can be lowered to support the backhoe when it is pulled . .

11-55. . . . to the loader use position.

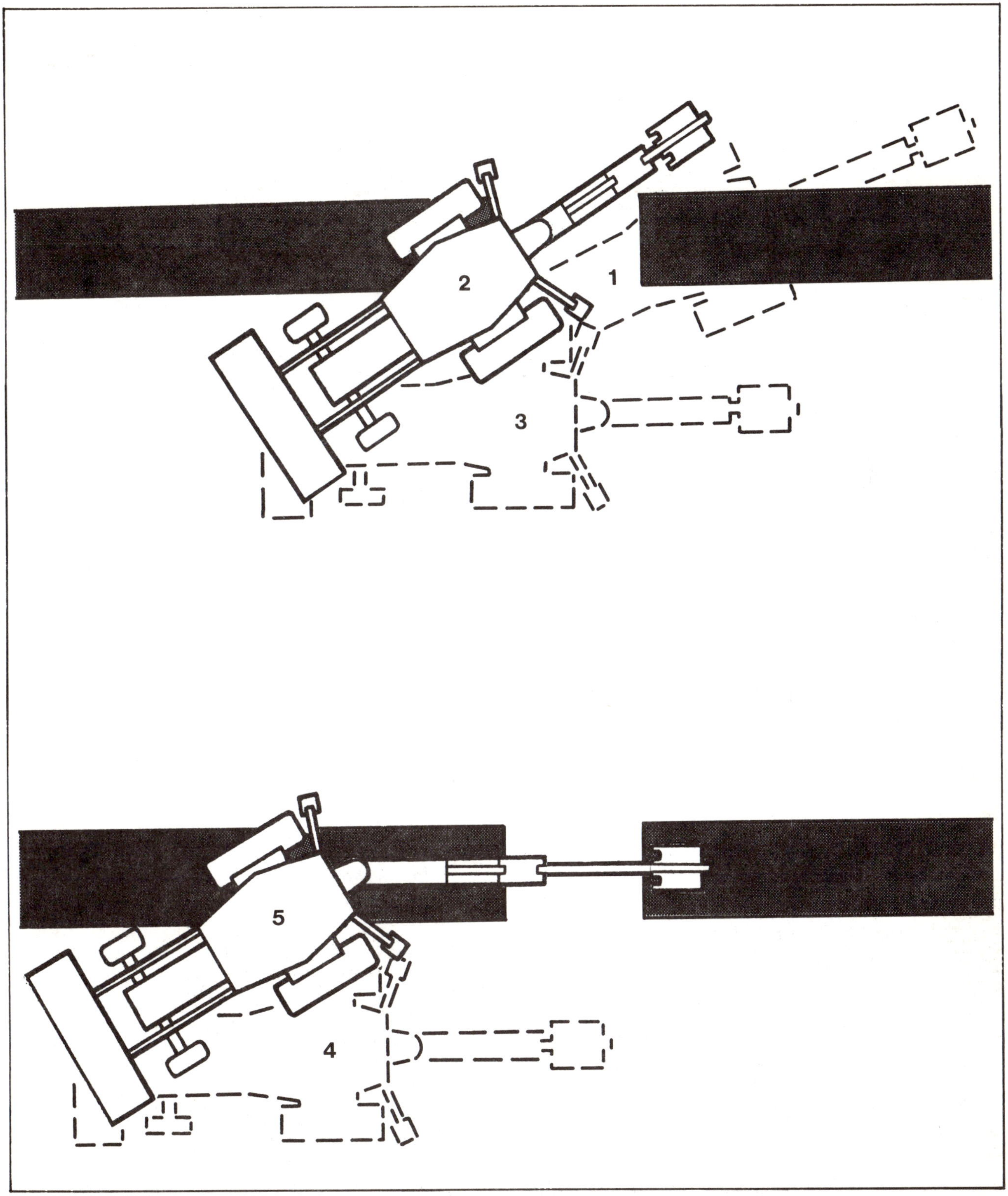

11-56. 1) When approaching the opposite trench, the machine will angle so that it can drive past the trench to the side. 2) This position is as far forward as possible so that the remaining area will be as short as possible. 3) Here the machine is pivoted away from the trench to the side. The loader bucket is used to stabilize the machine. 4) From position #3, the machine is moved forward to a point where it will pivot back across the trench in the exact desired position for the last section, (5).

12. LOADER TECHNIQUES

Introduction

Unlike most other types of heavy equipment which require the mastering of only one set of controls, the TLB consists of two machines in one—each with separate functions, controls, and required operating skills. Sometimes operators tend to ignore their loader skills while concentrating on backhoe skills, which is unfortunate because use of the loader is just as important as use of the backhoe. In on-the-job situations, operators who have made it their goal to master **total machine usage** will quickly stand out as being more valuable. When this type of operator is at the controls, the TLB becomes possibly the most useful all-around excavation machine in general construction.

Principles of Operation

The loader is basically simpler to operate and easier to get accustomed to than the backhoe. The hydraulics which control the lift arms and bucket position are also simpler to operate than the four backhoe controls. The hydraulics of the loader, however, provide only the **breakout force** of the bucket (Fig. 12-1), and the **lifting force** supplied by the two lift cylinders (Fig. 12-2). The real power that determines the usefulness of the loader bucket comes from the rear wheels. As a result, the amount of power available at the bucket is determined, in large part, by the amount of **traction** available in the area.

Traction

There are many forces and conditions which determine how much traction will be provided for the rear wheels. The most basic factor affecting traction is the amount of weight on the wheels, with traction increasing as the weight is increased.

As a forethought to this section, we should be reminded of Newton's law which says that for every action, there is an equal and opposite reaction. In any case where movement or forces are involved, there will be an effect on the machine and often, traction will be affected.

Driving forces. When the driving forces, or torque of the engine, is transferred into the ground, traction can either be increased or decreased. Fig. 12-3 illustrates how torque affects the machine when it is moving forward, while Fig. 12-4 shows the effect when traveling in reverse.

Hydraulic forces. Fig. 12-5 demonstrates that when pressure is applied to the lift of the loader, the weight on the rear wheels—and therefore the traction—is reduced.

The opposite is true when down pressure is applied to the loader—the weight of the rear wheels is increased, as well as the available traction (Fig. 12-6).

Load weight. The weight of the load being carried in the loader bucket will also reduce traction on the rear wheels (Fig. 12-7).

Ground conditions. Surface conditions on the job site can vary from a desert-dry dust bowl to a muddy, soupy mess. Also, soil conditions can range from hard-packed clay to loose sand or gravel. Depending on all these variables, the TLB can either be an effective excavating machine, or it can become awkward, cumbersome, or simply stuck. Because these conditions have a direct impact on TLB performance, the operator must monitor them, and be able to predict, and prevent, potential problems or dangerous situations. Optimum ground conditions that may have provided excellent traction can become dangerous and even deadly after a rain shower, for example.

Tire pressure. One easy way of increasing traction in soft or sandy conditions is to lower the air pressure in the rear tires. This can improve maneuverability greatly, but the operator should take care to refill the tires before using the machine on hard surfaces.

Stability

Slopes. As discussed in previous chapters, the TLB can often be an awkward machine to maneuver. When slopes are involved, this awkwardness can become dangerous. Generally speaking, the operator should avoid slope work whenever possible. When the TLB must be operated on a slope, the weight distribution should be adjusted by either **altering the dimensions** or filling the loader bucket. Once again, **do not** attempt any move you are unsure about.

Load height. Probably the most basic rule, whenever operating the loader, is to **carry the load as low as possible.** A filled loader bucket, carried low, will stabilize the machine (except when traveling forward down a hill). As the load is raised, stability decreases rapidly. The simple solution would seem to be to always keep the bucket positioned close to the ground—except that one of the loader's most common tasks is the loading of trucks, which requires raising the bucket high above the ground. As long as

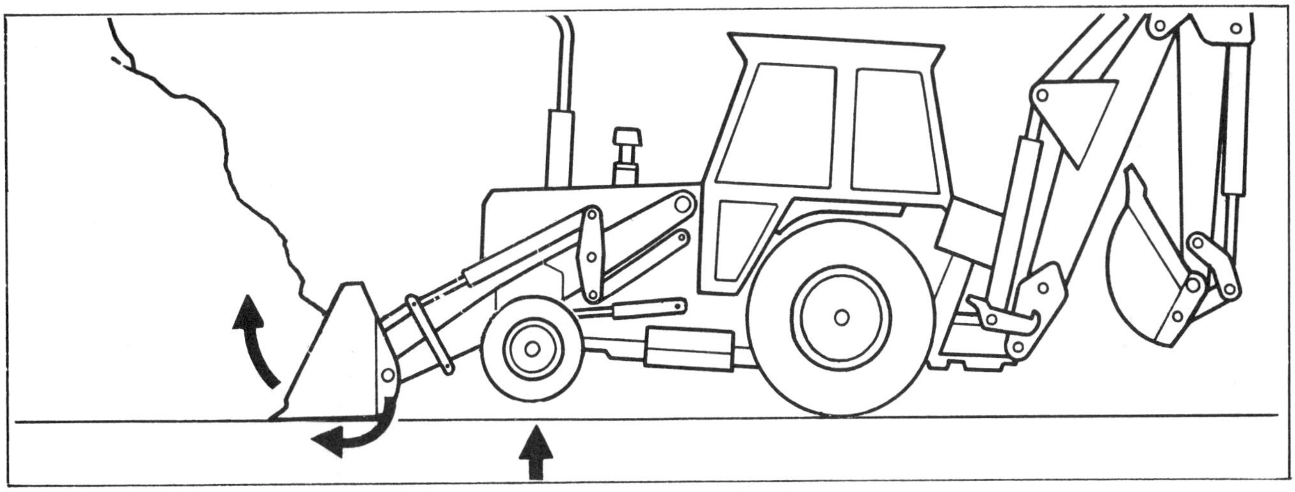

12-1. BREAK-OUT FORCE: Power is created through the bucket control hydraulics.

12-2. LOADER LIFT FORCES: A demonstration of the hydraulics which lift and lower the bucket.

12-3. **FORWARD DRIVING TORQUE** will remove weight from the front wheels and add weight to the rear wheels.

12-4. **REVERSE DRIVING TORQUE** will remove weight from the rear wheels and add to the front. This effect is greater when the machine is operated in low gear.

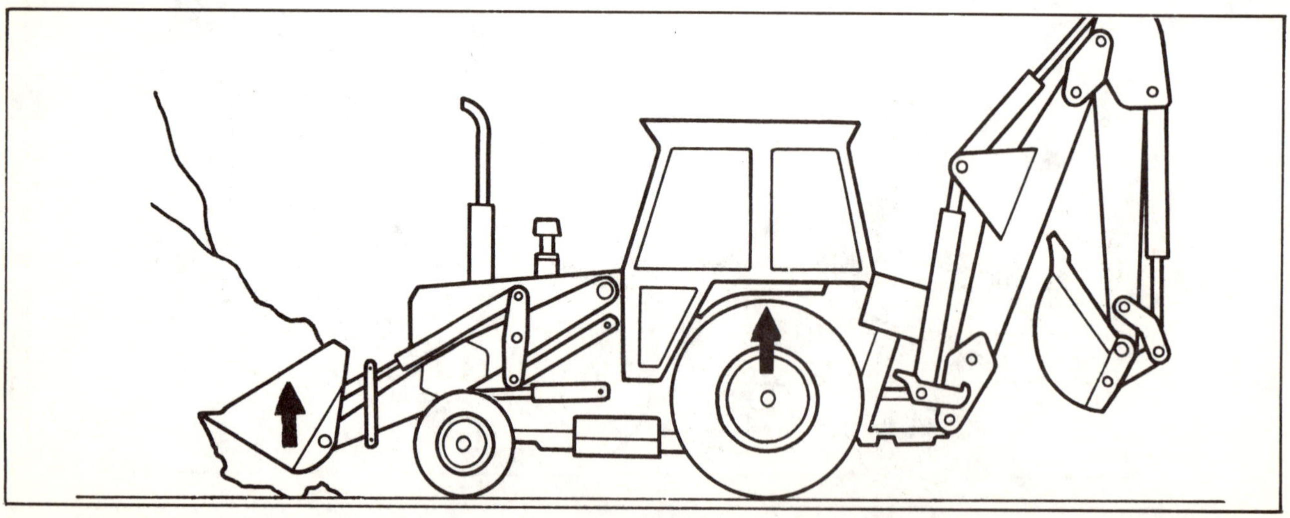

12-5. **LOADER LIFT**: When using the hydraulics to lift in a pile, the weight on the rear wheels will be reduced, as well as traction.

ground conditions are stable and level, there won't be a problem. If the area is sloped to any degree, or is uneven, bumpy or rutted, the truck-loading procedure can become dangerous and shouldn't be attempted unless the necessary corrections can be made. If the entire area is slightly sloped, the most stable approach to the truck is one shown in Fig. 12-8. Figs. 12-9 and 12-10, on the other hand, illustrate very dangerous and possibly destructive approaches.

By reviewing the points highlighted in the discussion of traction and stability, it becomes clear that the effectiveness of the loader is limited, mainly due to weight considerations. As the operator becomes more familiar with the unit and what it can and cannot do, he will also learn how to work around the machine's limitations. He will be able to complete the job using only those techniques that work well for the TLB.

Filling the Bucket

Nearly all tasks involving the loader require repeated filling and dumping of the bucket. The ability of the operator to get full buckets without wasting time will determine the effectiveness of many operations.

Back-ups. As mentioned in other chapters, providing a stable back-up is the single most important factor when loading either the loader or the backhoe bucket. By understanding this concept, the operator can often provide a back-up simply by the angle at which he "attacks" the pile, and he can maximize his efficiency by stockpiling in an area where a back-up is available, for example.

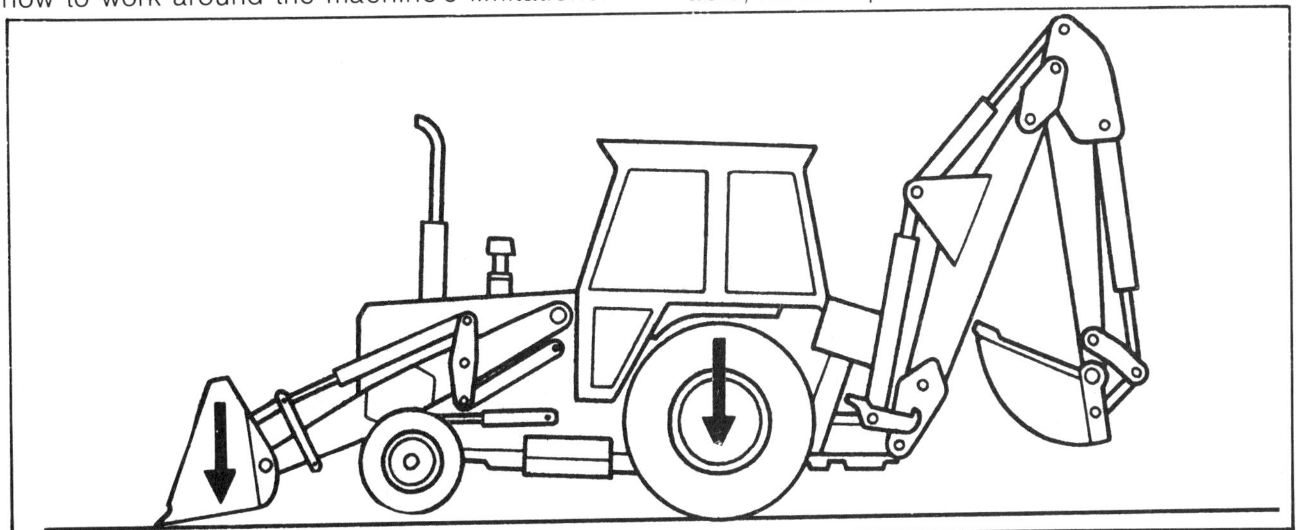

12-6. LOADER DOWN PRESSURE: When down pressure is applied to the loader weight and traction on the rear wheels is increased.

12-7. LOAD WEIGHT: When carrying a load in the loader bucket weight and traction on the rear wheels is reduced.

12-8. When the entire loading area is slightly sloped, the safest approach to the truck is the one shown here. The machine will be stable during both the approach and the dumping stages of the loading process. After the load is dumped, the front of the machine may tend to lift. The TLB should then be carefully backed up and the loader lowered to increase stability.

12-9. *Never* load a truck on a downhill slope, as TLB control could easily be lost.

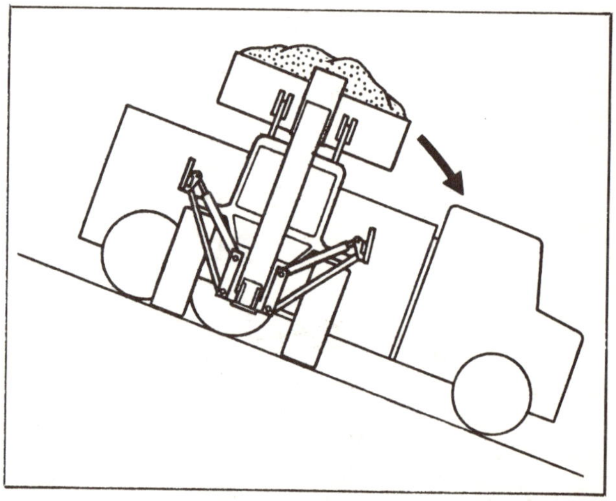

12-10. *Never* load on a side slope. Raising the bucket under these conditions will cause a rollover.

Corners of the bucket. In order to get full, even loads in the bucket, **the corners of the bucket must contact the pile at the same time.** If one side of the bucket contacts the pile at an angle, the bucket will not be filled to capacity.

First, the operator must select the area of the pile he wishes to "address" with the bucket. Then he should take note of the points at which the bucket will make contact, and the angle of approach should be one that ensures that the corners of the bucket will contact the pile at the same time. Fig. 12-11 illustrates four different situations and the correct approach for each.

Soft Material

Fig. 12-12. Enter the pile with the bottom of the bucket flat on the ground, applying as much power as possible without spinning the tires.

Fig. 12-13. The operator continues to apply as much power as possible, again without spinning the tires. The bucket is tipped back slightly and raised, pushing a growing pile ahead of the bucket.

Fig. 12-14. Here the bucket is tipped down slightly, the front tires are lifted, and slight pressure is applied to the bucket. With the greater traction available in this position, the bucket will penetrate the pile as deeply as possible.

Fig. 12-15. At this point a large pile has accumulated directly ahead of the bucket.

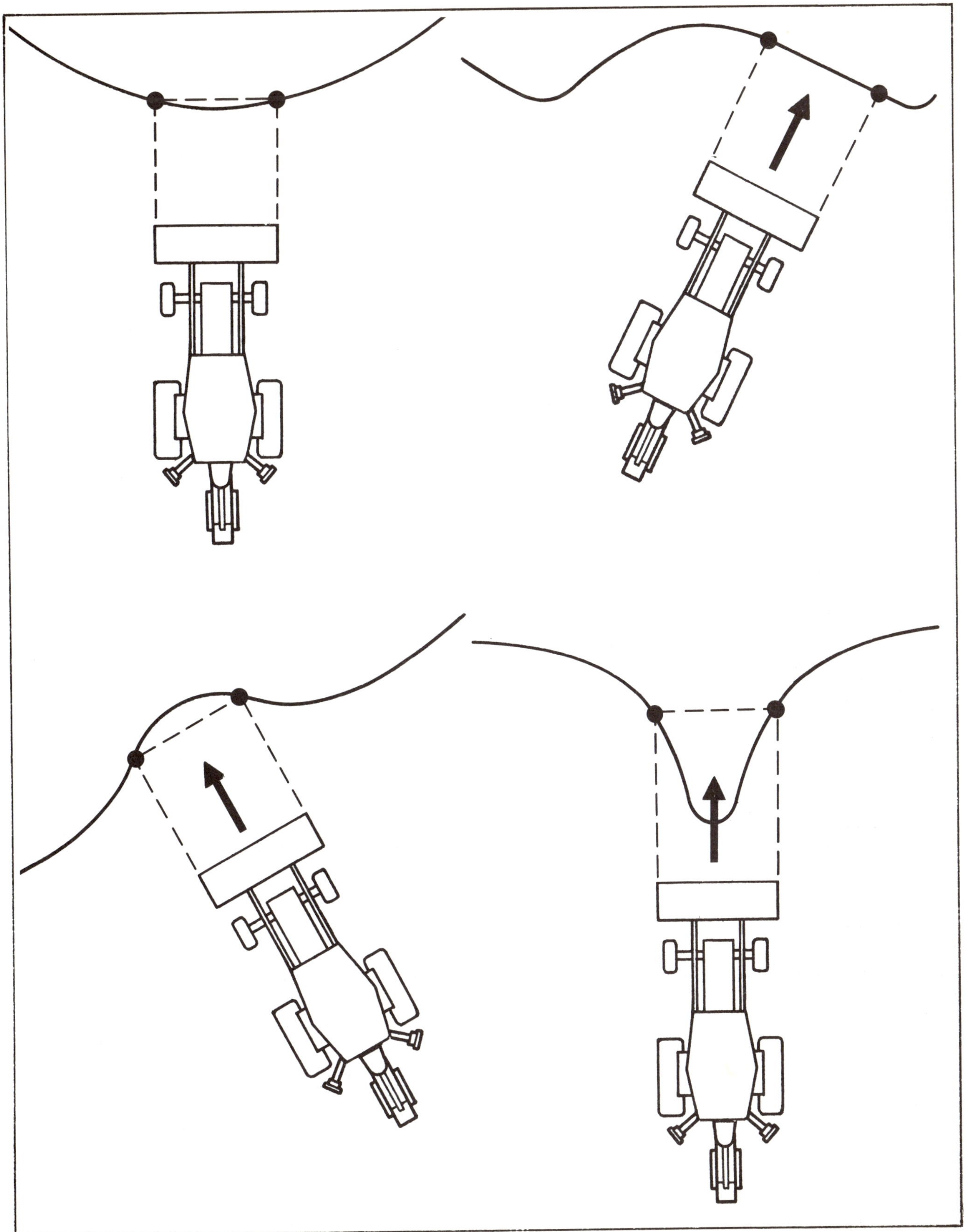

12-11. Regardless of the contour of the pile, the corners of the bucket should contact it at the same time.

12-12

12-13

12-14

12-15

12-16

12-17

12-18

Fig. 12-16. During this process, rolling the bucket down will push the dirt into the rear of the bucket. In order to quickly and completely fill the bucket, several rolling, up-and-down motions with the bucket will work well.

Fig. 12-17. The rolling motion described in the previous photo works so well, in fact, that the dirt is actually being pushed over the bucket.

Fig. 12-18. The goal of all loading operations —a full bucket.

When beginning to work a new pile of soft material, it is even quicker to take the first bucket from high in the pile. With the bucket tilted slightly down, the bucket will be forced down and deeply into the pile. Then a full bucket can be achieved by rolling the bucket upward (Fig. 12-19A).

The next pass will also be easier, since the weight has been removed from the lower material (Fig. 12-19B).

Hard Material

The loader is much less effective when attempting to load in hard ground conditions. It really doesn't have the traction capabilities or the breakout force necessary to handle such materials.

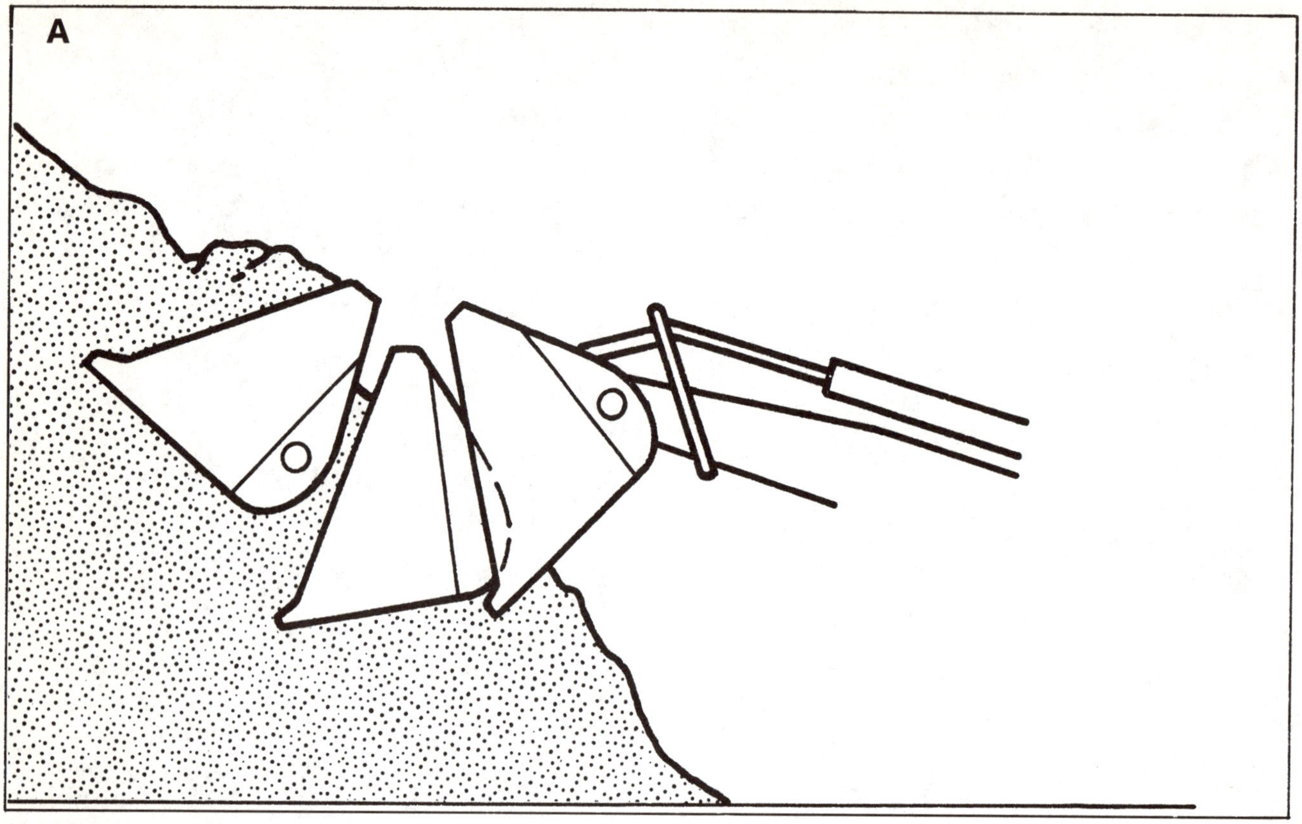

12-19 A). When loading from large, soft piles start from high in the pile. Drive the bucket down slightly, and deep into the pile. B). With the extra weight removed the second bucket will fill easily.

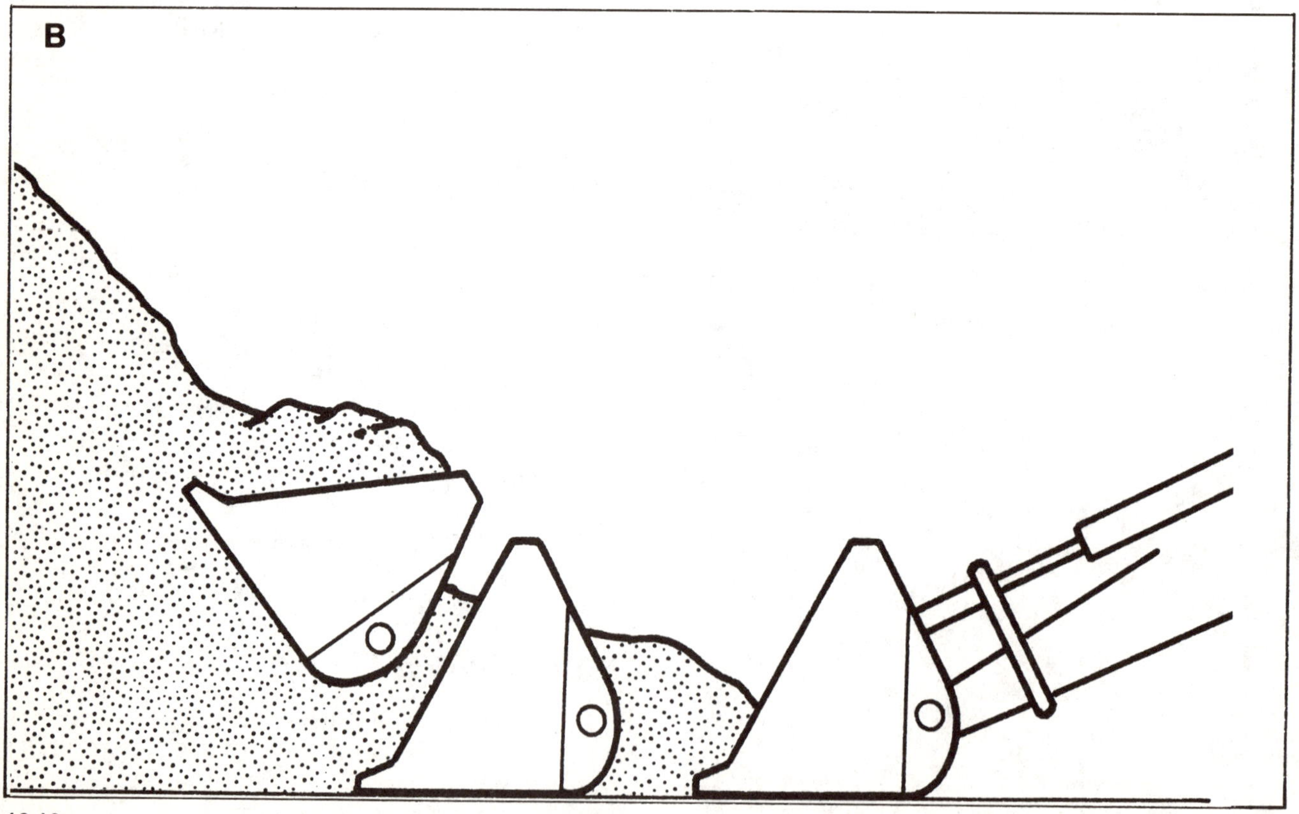

12-19

12-21

If the working area provides good traction, then it may be possible to penetrate the material. If the traction is poor, or if ground is soft, bucket penetration will usually not be possible.

The best approach, when dealing with tough ground conditions, is to use the driving forces in combination with the hydraulics to make a scraping motion similar to the one shown in Fig. 12-20. If this method is ineffective, then the hoe can be used to loosen the material, making the loader's work easier (Fig. 12-21).

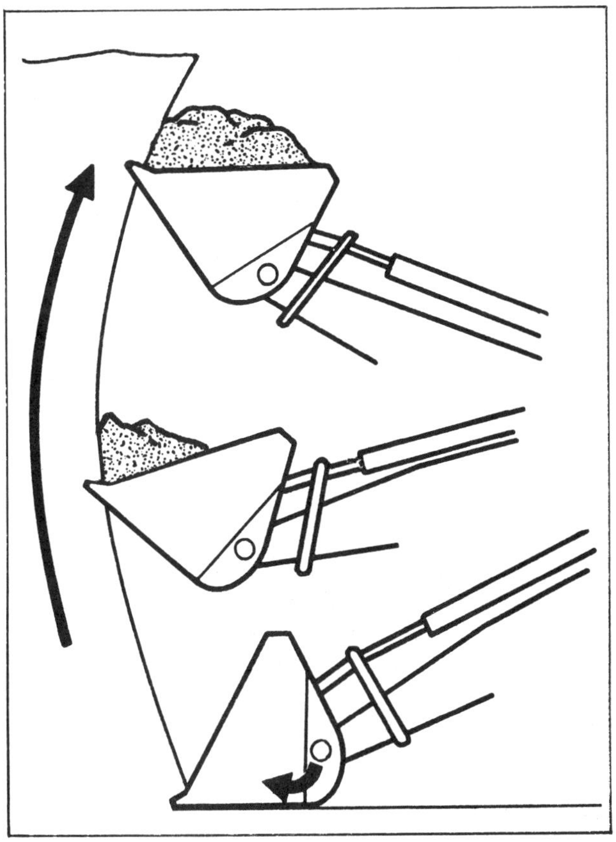

12-20

Loading Trucks

When the task of loading material onto trucks is at hand, the operator must first evaluate the loading area. What position should the truck be in? What loading patterns will work best? If there is a wind present, what direction is it blowing in? Does the area provide adequate stability for loading operations?

In selecting the loading area, the operator should look for an area that will allow the trucks to drive through—that is, where they won't have to back up in order to drive out. When they have to back up the loading process can be very slow.

If conditions are dry and dusty, the operator should consider the wind direction. If possible, trucks should be set up, and loading patterns chosen, in such a way that the wind will be coming from the rear as the load is being dumped. This will keep dust out of the operator's eyes and make the entire operation more efficient, not to mention more comfortable. Whenever possible, the TLB operator should assist the truck driver in locating the truck correctly, depending on the loading pattern.

Truck loading considerations. During the loading process, the operator will need to keep

12-22

12-23

several extra factors in mind. One of these is preventing damage. There are many ways in which the truck may be damaged during the loading process, such as damage to the sideboards of the truck bed (Fig. 12-22), or dents in the bed itself, caused by striking the truck with the front of the tractor (Fig. 12-23).

Another consideration is how the load's weight will be distributed when filling the truck. An improperly-loaded truck can become unstable when driven and could overturn on a curve or if the wheels go off the road onto the shoulder. Because of the height and reach limitations of the TLB, the load will tend to pile up on the near side of the truck, or at the rear, if the TLB is loading from that direction. To correct this, the load can be pushed across the truck bed as shown in Fig. 12-24. This will also help to fill the corners of the truck bed, ensuring a maximum load.

Stockpiling

As we've seen throughout the book, stockpiling is a common task—perhaps the most common loader procedure. Depending on the situation, anywhere from one-third to one-half of all jobs will require removal of spoil generated by the backhoe during excavation. Stockpiles are a common sight on many construction sites during the early phases of work. When footings are dug, spoil is often removed from the area where construction will take place, creating working space as well as a smooth working area. In instances where trenches are dug within the walls of a structure, the spoil will be replaced by sand or other materials that will ensure stability. And, of course, the stockpile will also provide fill materials needed for grading or back-filling.

It is easy to see that the crucial element when stockpiling is **access.** When selecting a stockpiling area, the operator should be mindful of the fact that the TLB will travel to and from the area many times. Free access is a prime consideration, and the chosen area should also not cause interference with other aspects of the job in progress.

Another helpful hint is to select a stockpiling area that contains available back-ups, as mentioned earlier. This is most helpful when the amount of material to be stockpiled is small, or if clean-up is to occur soon.

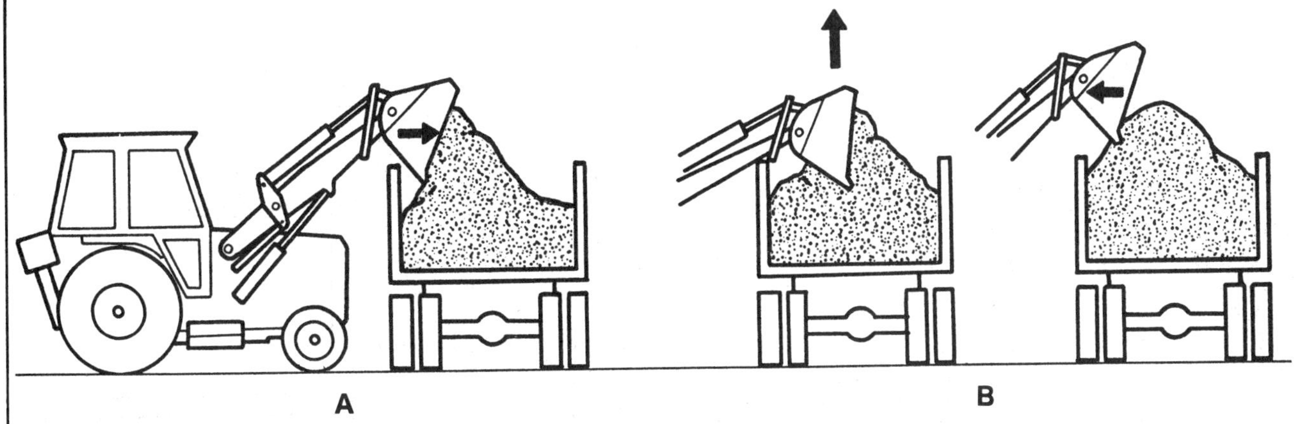

12-24. As the truck fills the load will need to be pushed across the truck bed in order to even the load. A) As the leading edge of the bucket passes the sideboard of the truck the bucket is rolled down quickly. B) The load is then pushed across the truck. By raising the bucket and backing up slowly the load will be distributed evenly across the bed. Be careful to watch the bucket, making sure it clears the sideboards of the truck bed, and also be careful not to strike the truck with the front of the tractor.

Ramp Stockpile

Although building a ramp stockpile is usually called for when larger types of heavy equipment are being used, it can be very useful to the TLB operator, particularly when stockpiling large, immeasurable quantities of material, and when the stockpiling area is limited. If the stockpiling area is adequate for the amount of material involved, the ramp method provides little advantage.

If, however, the available area is small, and a large amount of material is to be removed, then the ramp stockpile is clearly a better approach. By using this method, the pile can be built into a "two story" pile, and a much greater quantity of material can be stockpiled than by using the standard method.

Another advantage to using the ramp method is that the TLB's shuttle cycle will be the shortest and quickest possible for that particular job situation.

This method is not without some disadvantages and potential problems. If the ramp is too steep, the machine can become dangerously unstable, and steering control can be lost. To avoid this, the ramp must have a low degree of slope (Fig. 12-28).

Also, the "edges" of the ramp, and the "platform" area on top of the pile will be loose and will not support the machine's weight. It may appear to be safe and stable, but **it is not.** It is most important to keep the rear wheels away from the edges; if the rear wheels were to sink only a little at the edge, the weight distribution immediately shifts to that wheel, causing even more sinking.

To avoid these problems, the ramp should be built to a minimum of one and a half times the width of the machine. After this wide area is compacted and graded, the operator should keep the TLB in the center of the "roadway," away from the edges.

Another disadvantage of this technique becomes apparent when the stockpiled material is removed. The hundreds of passes over the material will compact it to the point where removal with the loader may be difficult. In this case, the pile may have to be loosened with the hoe before removing it with the loader.

Ramp building technique. Unlike the standard stockpiling technique, which is started at the farthest point required for the amount of material expected to be stockpiled, the ramp stockpile is started very close to the work area. Basically, the pile is started by building a ramp and adding material over the end (Fig. 12-29). As the ramp is formed and the material is added, the operator grades and smooths the ramp during the shuttling of material to the ramp area. After a load is dumped in the desired place the loader bucket is then used to back-drag the ramp as the machine returns to get another load.

TRUCK LOADING PATTERNS

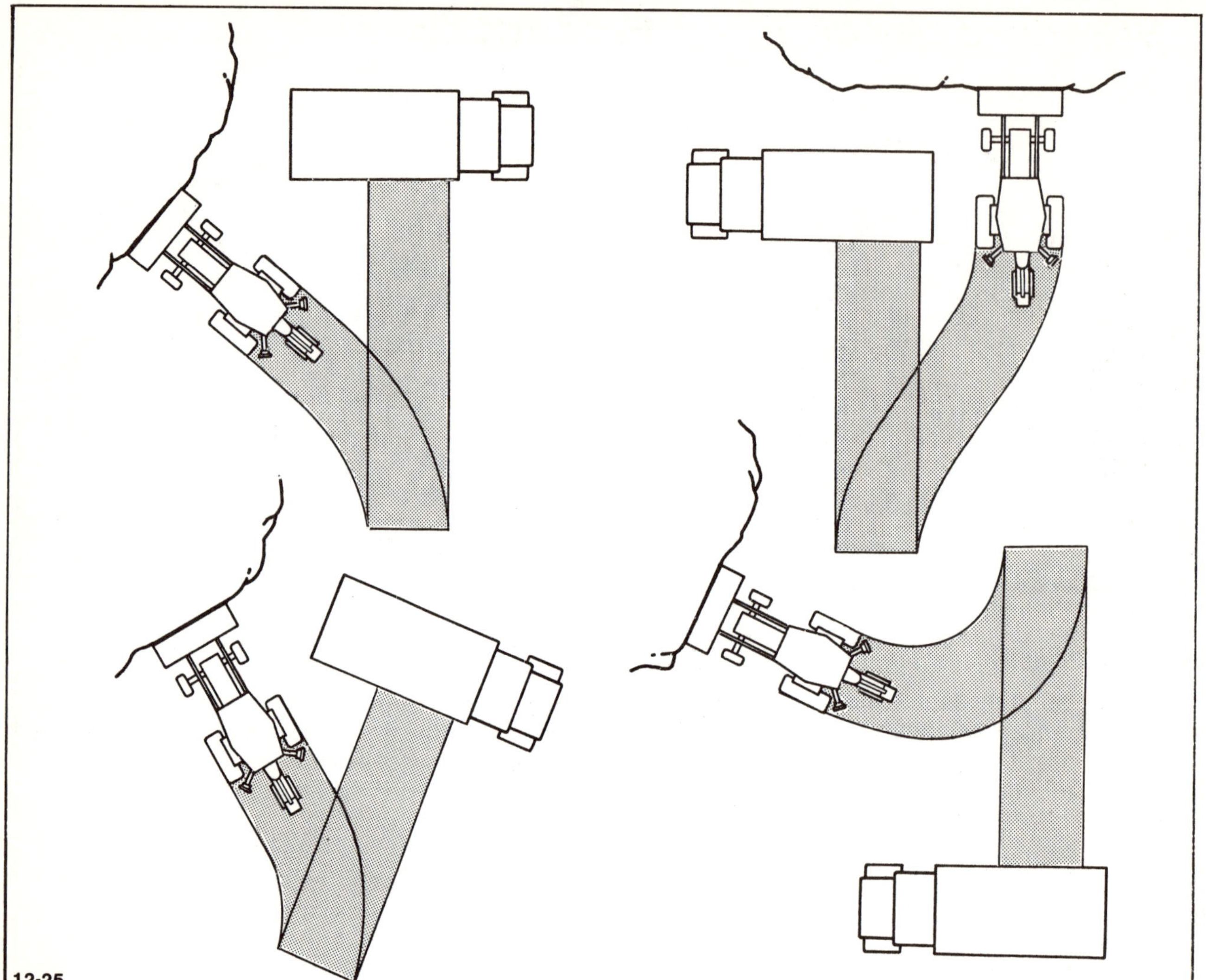

12-25

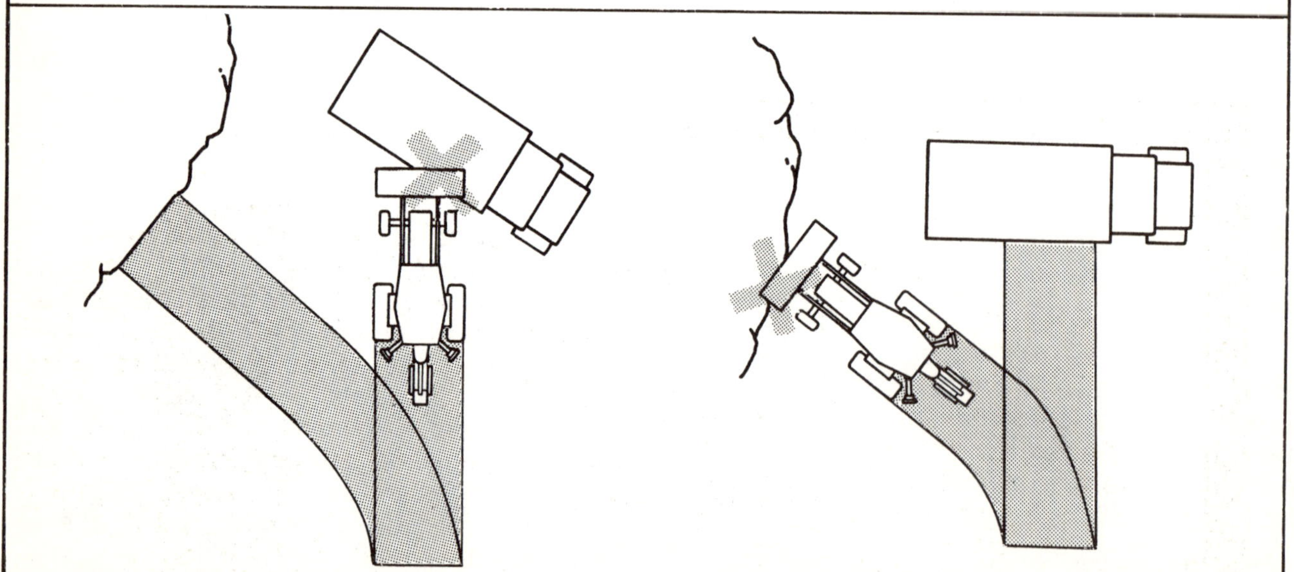

12-26. Always choose a loading pattern that will allow for the bucket to enter the pile and dump into the truck evenly.

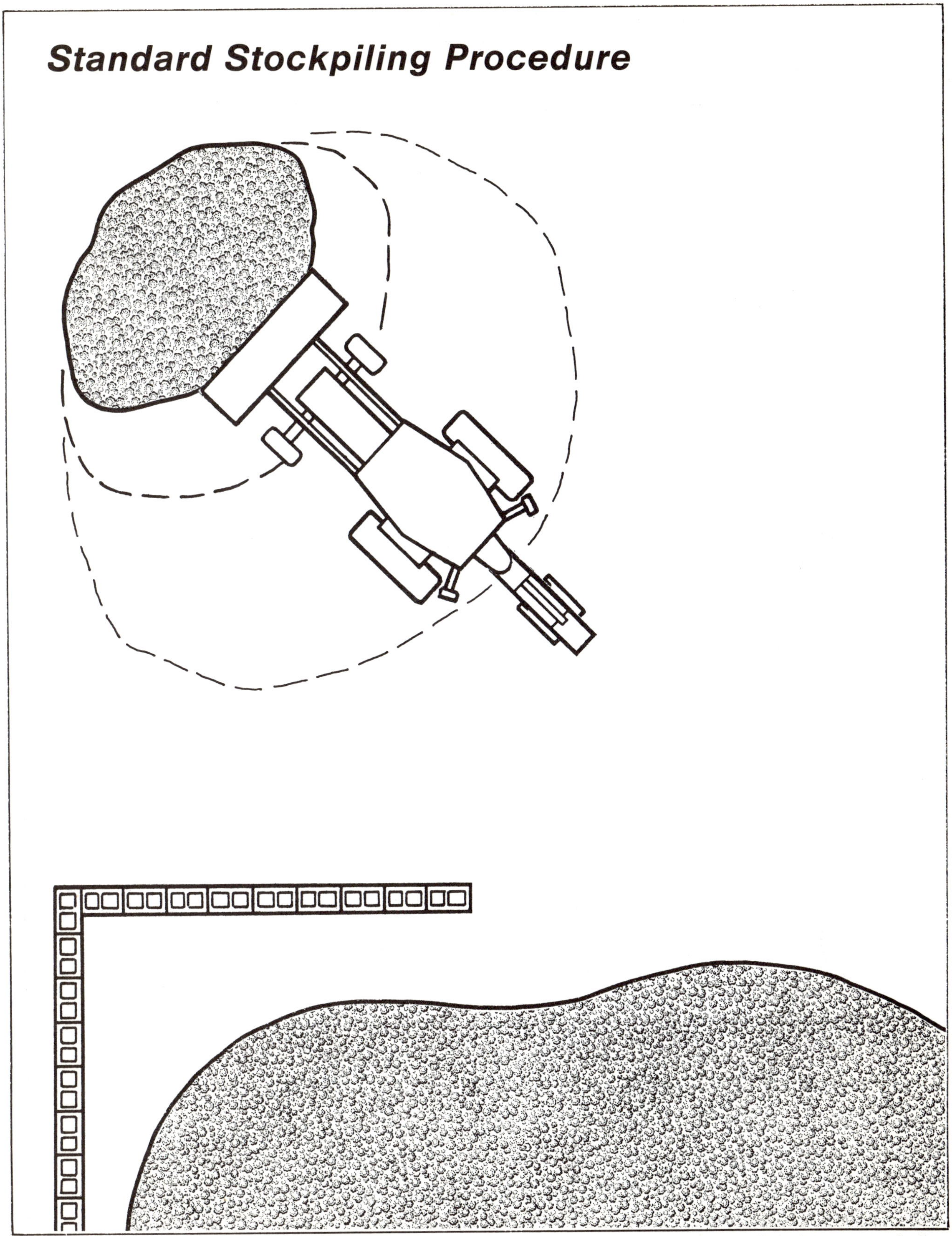

12-27. When using the standard stockpiling procedure, the pile is started at the farthest point available, or the point necessary for the area and the amount of material involved.

12-28. The ramp should be formed at a low angle to ensure stability during shuttle operations.

12-29. As the ramp is enlarged material is added over the end.

As the ramp grows, it should be raised to the desired height and gradually leveled off. Then the piling is continued with more material added over the end until it reaches the limits of the stockpiling area (Fig. 12-30). As material is added, the pile on the top will grow to the edges of the original pile, and then the "second story" pile will be made in the same manner as the standard stockpiling procedure (Fig. 12-31).

Related Techniques. Due to the usual lack of space found on most construction sites, the stockpile may have to be moved or condensed into the smallest area possible.

Moving small stockpiles short distances. It is not uncommon during the course of construction that stockpiles will need to be relocated, either to provide better access to the area, or to create space for a new excavation. The technique, which is illustated in Figs. 12-32 through 12-35, is also useful for back-filling trenches.

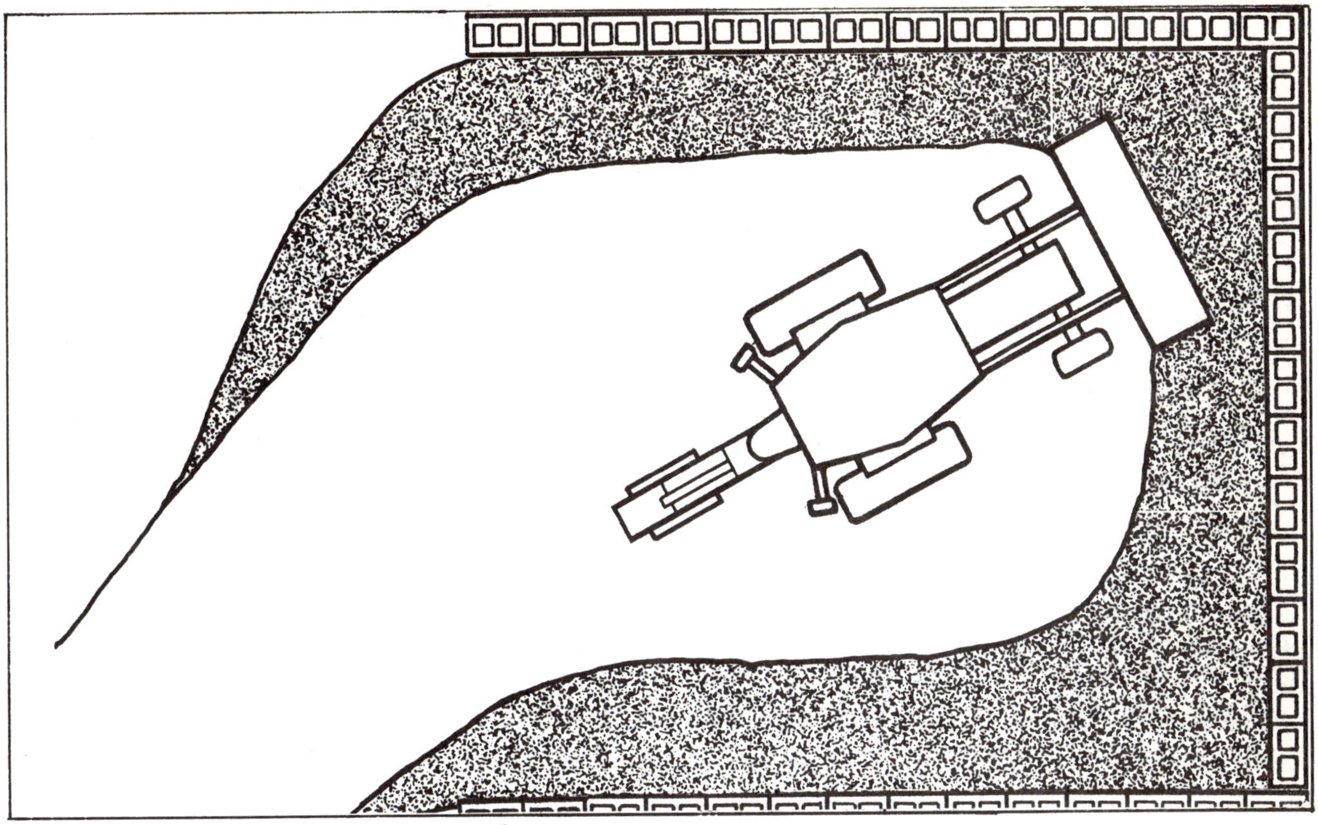

12-30. When the available stockpiling area is used . . .

12-31. . . . a "second story" is added to the pile.

12-32. The pile is entered with the bucket flat.

12-33. The bucket is raised as power is applied . . .

12-34. . . . until the top of the pile gives way . . .

12-35. . . . and the bucket is driven forward. From this point the process is repeated, starting with the bucket flat on the ground (as in Fig. 12-32).

Condensing the stockpile. During the general clean-up of the job site, gathering spoil from various excavations and piling it all into one or more stockpiles, the material may need to be consolidated or condensed into a single pile. By doing so, the job site is cleared of smaller spoil accumulations and the area becomes more accessible for the other trades on the job, which facilitates the entire construction process.

In another situation, material that has been trucked in may have to be condensed. When the trucks dump the material, it is usually piled low and spread over a large area. To save space the material should be pushed into one large pile as high as possible. The method used to accomplish this task is illustrated in Fig. 12-36.

Shuttle Operations

"Shuttle" is defined as "to move or go back and forth rapidly or frequently." And that is exactly what the TLB does during shuttle operations. Up to this point, we have divided the discussion on loader operation into several categories: principles of operation, including loading the bucket; various job tasks, such as loading trucks and stockpiling; and so on. The one element that combines all of these various operations, linking them together in a workable progression of TLB functions, is the shuttle operation. Shuttling binds the otherwise separate loader operations together into a smooth, efficient job.

Essentially, shuttling consists of moving buckets of materials from one point to another. This material will usually be dirt, but it may also

12-36. A large, flat pile may be condensed using the motion shown here.

12-37

be sand or gravel, other building materials, or any materials which are difficult to remove by hand.

Shuttling is also involved during clean-up of the job site, consolidating spoil piles into a central stockpile area, as noted earlier.

Another very common shuttle operation is the procedure of loading trucks. Actually, truck loading involves three basic elements of loader operation: filling the bucket, the actual loading of the truck, and—linking these two procedures—shuttling back and forth from the truck to the loading area.

Because shuttling is involved in nearly every loader job, it is easy to understand why safe and efficient shuttling is important to loader operation.

Efficiency. The greatest single thing that will ensure efficient shuttling is **a smooth, well-graded shuttle route.** Without a smooth route the TLB can bounce about wildly and control can easily be lost. The TLB is affected by bumps in the work area to a much greater degree than other types of heavy equipment due to its uneven weight distribution. If the operator does not smooth the area, the front tires can be kicked into the air by a bump when traveling with the loader bucket empty (Fig. 12-37). A rough shuttle area will also cause materials to fall out of the bucket along the way, creating even more bumps.

In order to keep the shuttle area smooth, the operator should back-drag the route when traveling back to get another bucket. Also, before leaving the pile area with a full bucket, the operator can give the bucket a couple of short, quick shakes. This will leave any loose excess material at the pile and will also settle the remaining material in the bucket. This can also be helpful when the shuttle path travels over paved surfaces that will require clean-up at the completion of the job.

Length of the shuttle. By examining the **efficiency** of any shuttle operation, we can see that it is determined by the amount of material moved within a given time period. Obviously, this means that a shorter shuttle run will be more efficient and productive over the course of the job. So, when planning the shuttle route, try to use the shortest possible path between the two endpoints, as well as one that provides for easy filling and dumping of the bucket.

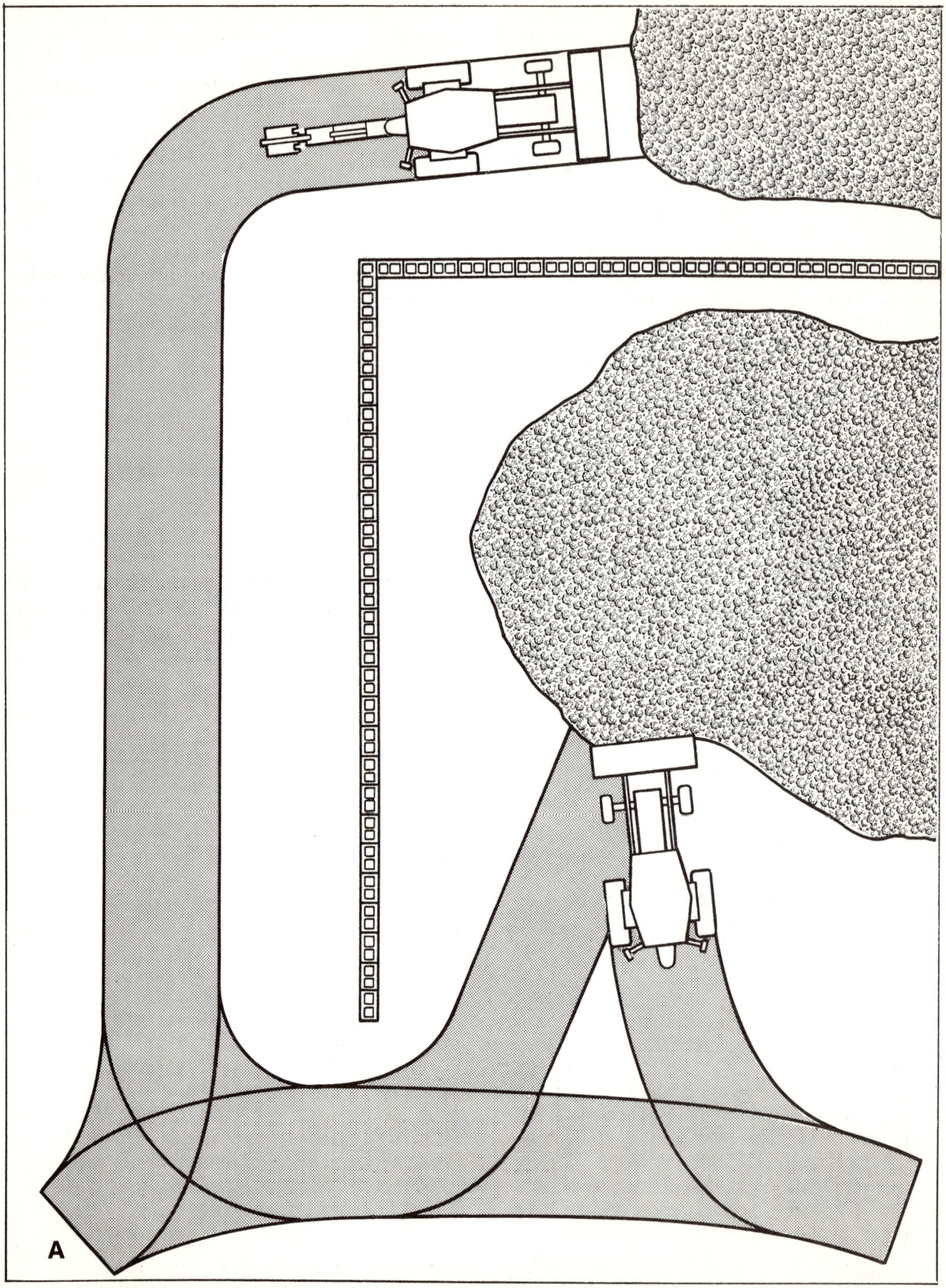

12-38. Shuttle operations will often stretch from one end of a job site to the other. Position "A" is detailed in Fig. 12-39.

Bucket volume. Here the length of the shuttle is again a determining factor. When the shuttle run is **short** (twenty to fifty feet from the point where the bucket is loaded to the point where it is dumped), less than full buckets are acceptable. The object of filling the bucket is to get as much as possible without wasting any time. Entering the pile a second time would surely be a waste of time.

When the shuttle run is **long** (100 feet or more), the operator should fill the bucket as completely as possible. In this situation, a second pass to completely fill the bucket is preferable to traveling the entire distance with a half-full bucket. When filling the bucket for a long shuttle run, follow this two-step procedure:

1) Enter the pile and fill the bucket as much as possible. This should require only one pass, but if the bucket is not completely full, take the extra time to fill it.

2) As the bucket is lifted from the pile, prior to backing away from it, shake the bucket quickly to remove excess material to keep the shuttle route clean. The actual shaking action is accomplished by rolling the bucket out slightly from the fully-curled position, and then quickly rolling it back against the bucket stop position.

By following this procedure, the bucket will be completely full and it will also be "well groomed" for the long shuttle run.

In order to better grasp the principles behind shuttle operations, we will examine a long, but not unusual, shuttle procedure. Fig. 12-38 illustrates a shuttle pattern that might be used on any common construction site. This pattern might be repeated several times, or hundreds of times over several hours, or even all day. Here are some important tips and pointers that will help the operator better cope with this type of work.

"Radar." One thing an operator can always expect on any job is a set of certain restrictions. Any object in the work area can be considered a restriction, such as grade stakes, building materials, cars and trucks parked on the site, fences and walls, trees, buildings, utility lines, building overhangs, overhead wires, etc.

On the first several trips along the shuttle route the operator will note where such restrictions could pose problems. Moving slowly and carefully through and around these hazards, and noting the clearances, will enable the operator to reposition the machine through the areas where obstacles pose a problem. After several passes the hazard areas will become part of the overall

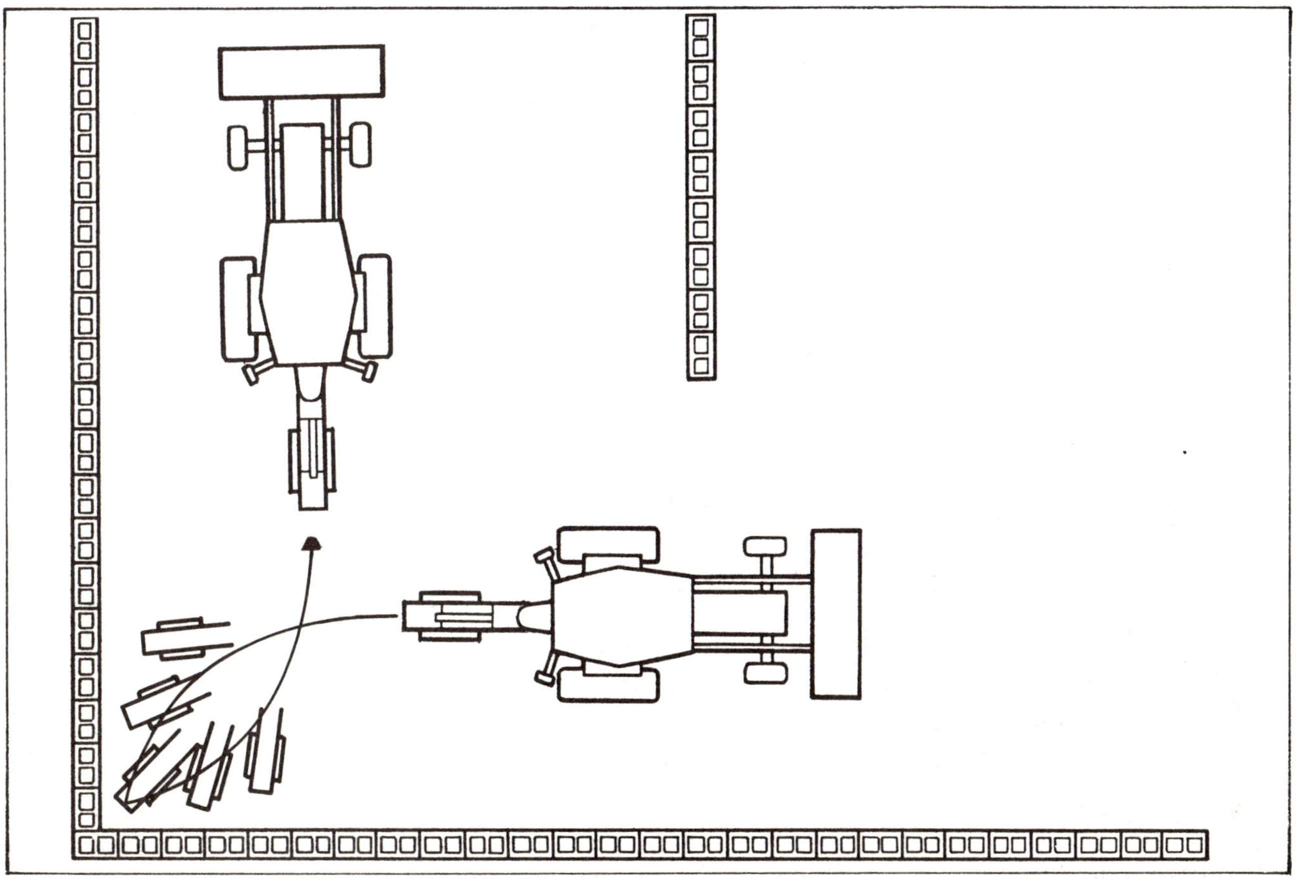

12-39. This illustration shows in detail how the operator can maneuver the TLB in tight situations using the "corner" of the work area.

shuttle cycle and the TLB will follow an exact route that will avoid them. As this procedure is repeated many times the operator will develop a sense of "radar" and will know his position at all times when traveling the shuttle path.

It is important to note that this "radar" system works only on inanimate objects and doesn't apply to people working on the site, moving vehicles, or other heavy equipment.

Use corners well. When space becomes a problem while traveling the job site, the "corners" of the work area can become useful when maneuvering the machine. As seen in position #A in Fig. 12-38, the operator uses the corner to change the machine's direction. Fig. 12-39 gives a more detailed view of the motion used in this maneuver.

Direction of travel. When traveling the shuttle path the first few times, the operator will develop a pattern that will be repeated many times. If there are any slopes within the shuttle path the operator should decide which direction (forward or reverse) he should drive the machine in order to maintain the greatest control over it.

Safety

Traveling in reverse. Because of the size and position of the backhoe attachment, the vision of the operator can sometimes be obstructed (Fig. 12-40). To overcome this problem the operator must turn his body to one side and lean slightly in the same direction. Then, after surveying the situation on that side of the machine, he shifts around and does the same thing on the other side (Fig. 12-41). By following this procedure the operator will get the full view of what is behind the machine.

People. Extra attention will be required any time there are workers in the shuttle area. People often get so involved in whatever they are doing that they don't think about what is going on around them. **Do not assume** that people will not walk into the path of the machine. Watch out for them and be prepared to stop or take other action to avoid accidents or injuries.

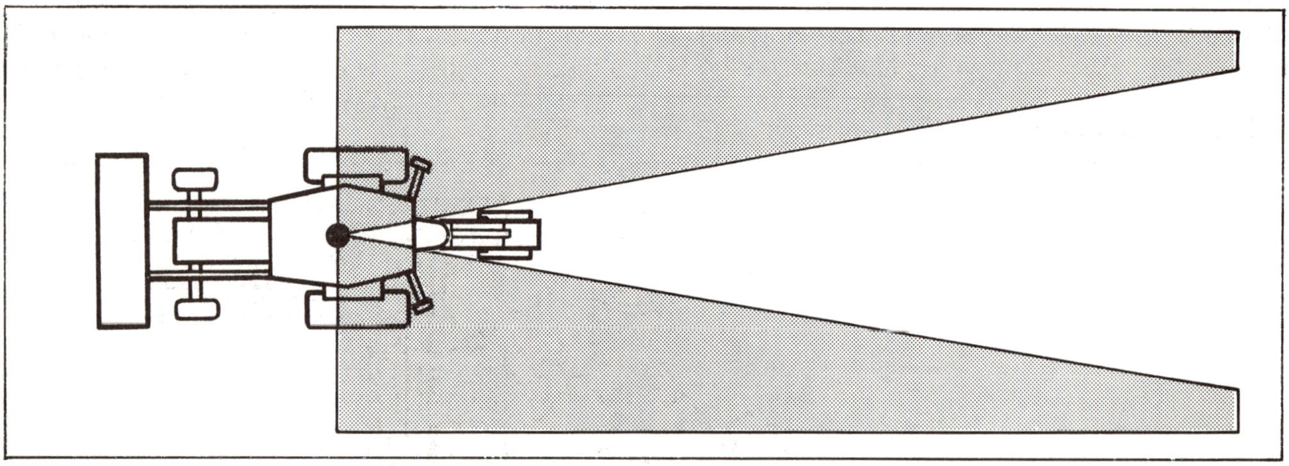

12-40

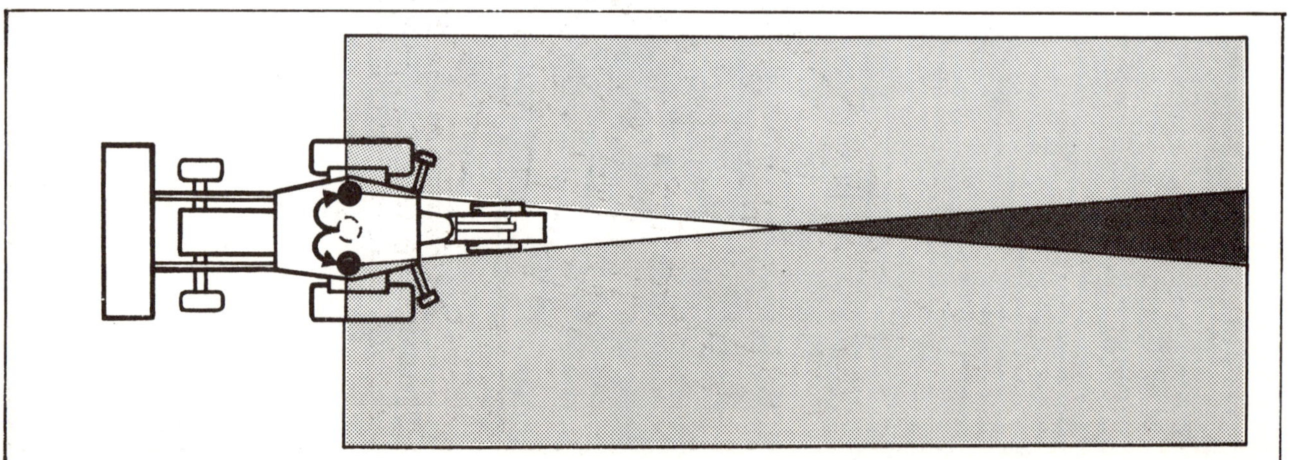

12-41

Related Techniques

Road building and leveling. This technique applies mainly to those times when ramps, driveways and shuttle paths must be built on the job site. The procedure is similar to the cut-and-fill technique described in the chapter on ***Grading and Compaction,*** particularly the section on "starting the cut" (see Figs. 10-3, 10-4 and 10-5).

The key to easy road leveling is to provide and maintain a level area for the TLB to work from. Once the machine is positioned properly in a level area the grading and leveling can be accomplished correctly because the bucket will cut at the desired angle.

As the cut-and-fill sequence takes place, the fill side should initially be graded higher than the cut side. This will allow for compaction and will ensure the procedure is carried out safely. After the new roadway has been thoroughly compacted it can be regraded and smoothed until level, if desired.

Another safety reminder when building a roadway: always try to widen the road to the point where it is at least one and a half times the width of the TLB. This will allow easy maneuvering and reduces the risk of running the machine off the edges.

Circular grading patterns. Although this technique may not be called for frequently, it can be very useful when the operator faces a grading job in a tight situation, such as back-filling and grading a trench that is adjacent to a fixed structure, such as a wall or a house.

The basic procedure involves positioning the TLB at the most comfortable angle relative to the trench, with one corner of the bucket contacting the pile of back-fill material. Then, as the operator backs up, away from the trench, he engages the brake on the rear wheel farthest from the trench so that the TLB begins to move in an arc. By carefully engaging and releasing the brake the operator can make the front bucket move in a circular pattern, pushing back-fill material into the trench in a sideways motion, either left-to-right or right-to-left, whichever the case may be. Figs. 12-42, 12-43 and 12-44 illustrate the correct procedure and the desired results.

The most common applications of this technique include: back-filling or grading next to buildings and walls; grading close to houses or delicate surfaces; back-filling against basement walls, where keeping the machine's weight back away from the structure is important; grading small or narrow areas that won't support the machine's weight; or when working in small areas where grading may be otherwise restricted or inhibited.

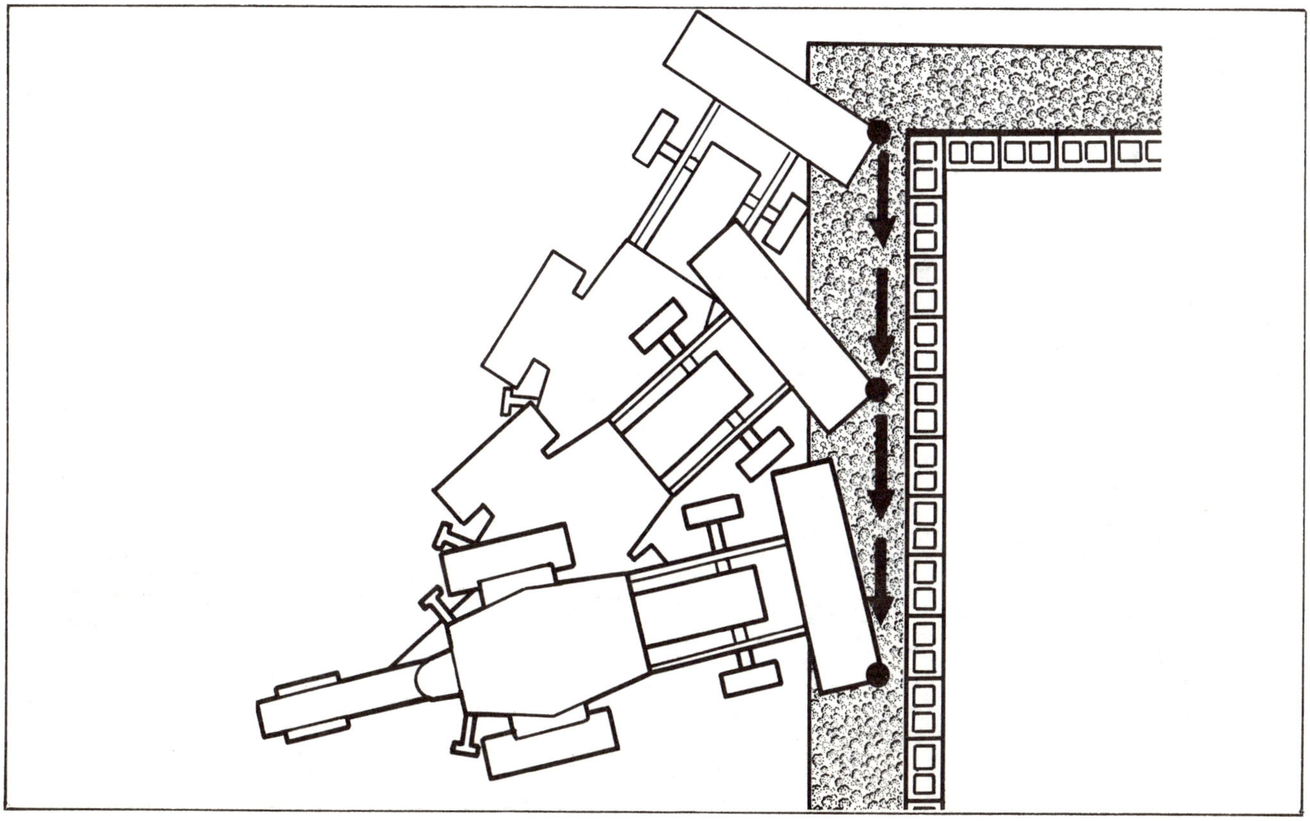

12-42. When grading in a circular pattern the loader bucket is located far from the weight of the machine, and can be controlled easily for grading next to delicate structures.

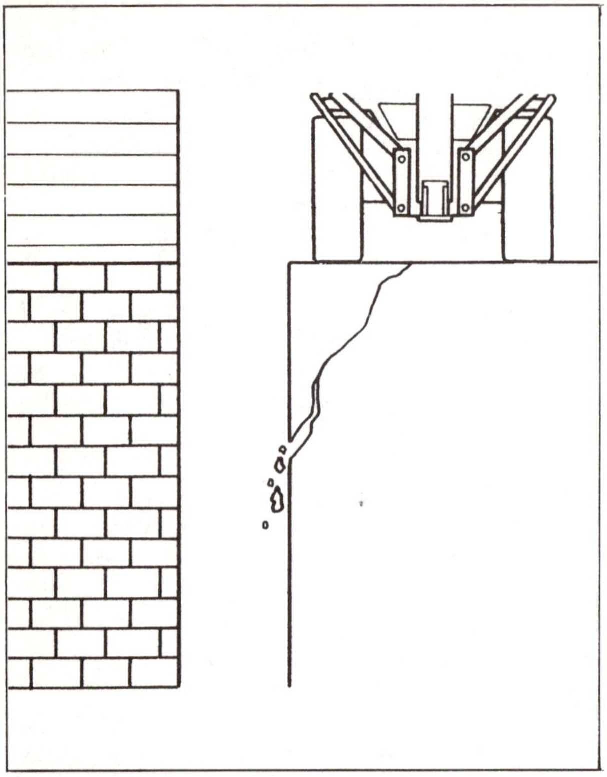

12-43. Positioning the TLB next to the excavation, as shown here, would be unsafe.

12-44. Positioned at an angle the weight will be supported on stable ground.

12-45. When cleaning trench sides the TLB is positioned at about 45 degrees relative to the trench, with at least half of the bucket over the edge of the trench.

12-46. By applying the left brake the bucket will be slid quickly across and away from the trench . . .

12-47. . . . resulting in a clean trench side and little or no material pushed back into the excavation.

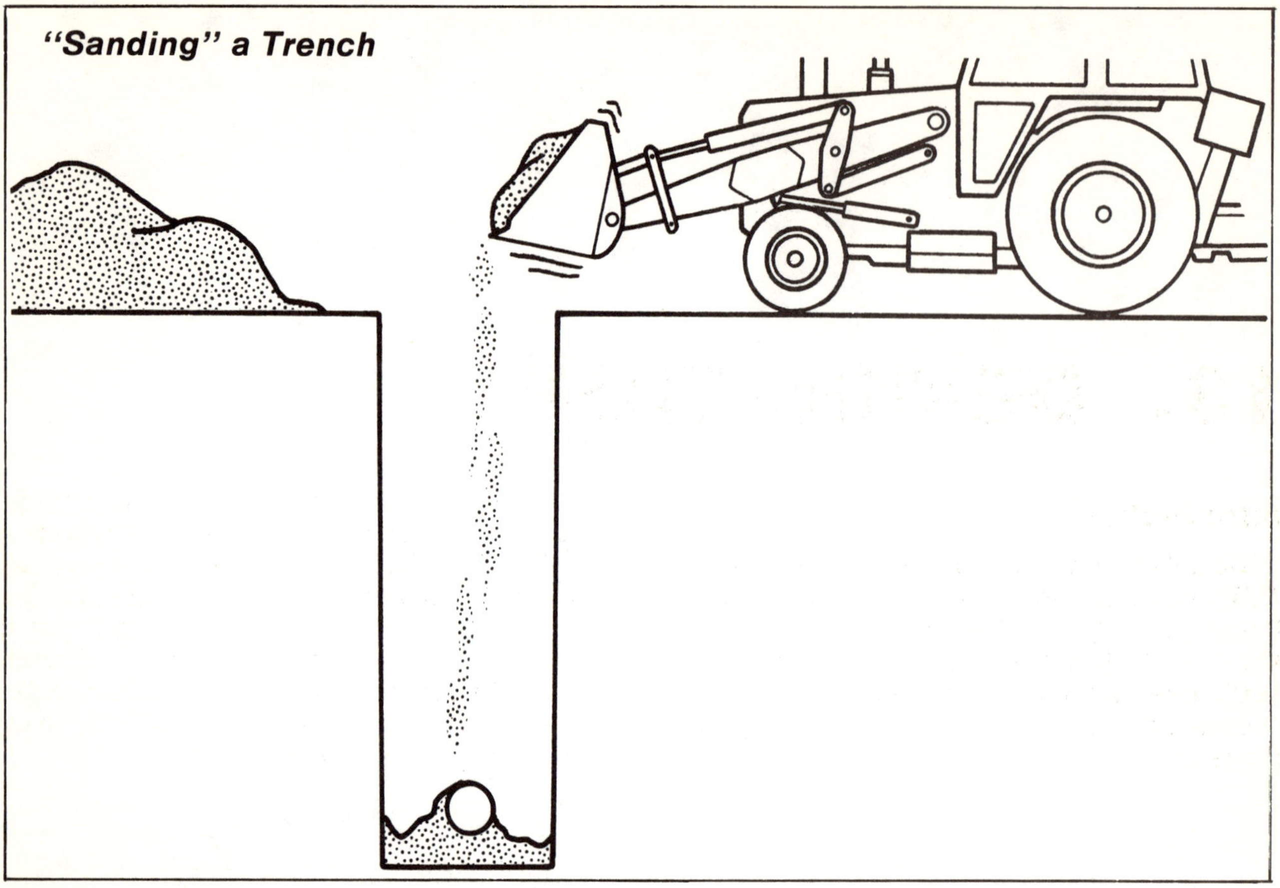

12-48. When "sanding"a trench for a newly-laid pipe, sand must be carefully shaken from the bucket so it is distributed evenly, falling like rain into the trench.

13. DEMOLITION

Introduction

Up until now, we have focused primarily on the TLB's many applications in construction, and the preparatory work leading up to actual construction. In this chapter, we will discuss the TLB's role in demolition of existing structures such as sidewalks, driveways, walls, parking lots, curbs and gutters.

Breaking and removal of concrete, asphalt, etc. is a very specialized area of TLB operation; while nearly anyone can break up and remove concrete, it takes technique, experience and common sense to do it efficiently. This type of work often calls for the operator to use common sense to find solutions to unusual problems. The TLB is, however, an excellent choice for many small breaking and removal jobs.

13-1. The TLB can be extremely effective when used for breaking and removal of many types of common structures, as shown here.

Preparation and Planning

Like any other type of TLB work, the demolition job must be planned, step by step, and the operator must have a clear job sequence in mind. Before beginning, the operator should walk the job site to get a clear idea of the scope of the work to be performed. There are many elements common to most demolition jobs, and they are inter-related. The concepts which the operator must consider include:

- **Efficiency of job completion,** which is determined by how many times the material being removed is handled.
- When material is to be handled two or more times, there are several **stockpiling** considerations to be made.
- As in any loading or material handling operation, providing an adequate **back-up** ensures efficiency.
- In many job situations, **separation of different materials** is an important consideration.
- **Clean-up** is either carried out at the end of the job, or planned in stages as work progresses.

Let's examine each of these factors individually.

Efficiency of Job Completion

When removing material from a demolition site, efficiency is determined by how many times the material is handled. The best method, obviously, is to handle the material only once. That is, it should be loaded onto trucks immediately as it is removed from the site. This is accomplished with the backhoe, working along with a truck that is positioned to facilitate loading.

In many instances, using this method will mean that the truck will have to be moved often as work progresses. This approach may seem slow and difficult, especially to beginning operators. However, if we examine the alternative—first removing the material to a stockpile, breaking it into pieces small enough for the loader to handle, and **then** loading the truck—it becomes clear that spending a little extra time loading the material as it is removed will really be quicker in the long run.

The key to the one-step method is **patience,** because the operator will often have trouble positioning the pieces in the bucket. Also, the operator shouldn't waste time trying to get an extra piece onto the load. Productivity is determined by how much material is loaded within a given period of time. By trying to jockey one more chunk into the bucket, however, the operator will usually end up losing the pieces he has already positioned correctly. The actual loading techniques used in this procedure are explained later (see ***Loading with the Backhoe***).

There are many situations where one-step removal won't be possible. For instance, space sometimes won't allow the truck to be positioned alongside the TLB. In other cases, material removal will be so time-consuming that stockpiling it for later loading may be the best approach.

The main point here is to avoid stockpiling the material and load it immediately. Any time stockpiling is involved, it means that material will have to be handled at least twice, or even three times. When planning job efficiency, plan to handle the material the fewest possible number of times.

Stockpiling

When stockpiling becomes necessary, the main considerations are how the material will eventually be loaded out, and how to stockpile in such a way that loading will proceed quickly and easily. There are several things to consider when choosing the stockpiling method:

- **The area.** When deciding **where** material will be placed, the area chosen must be large enough and accessible enough to permit loading and maneuvering of the trucks.
- **The size of the pieces.** Generally, when the loader is used to load concrete, the larger the pieces are, the more difficult the loading process will be. One rule of thumb when loading concrete with a machine of this size is that the pieces should be small enough to move by hand. This is accomplished by breaking the concrete into pieces of the desired size **as it is removed.** This will ensure full buckets when loading, and will also make clean-up easier.
- **Keeping the pile low.** Because of the irregular shape and great weight of the concrete pieces, piling them high tends to result in friction, causing the pieces to become "tangled" in the pile. Loading from a highly stacked pile is difficult and frustrating. To avoid this, keep the

stockpile low—no more than two feet high—and spread out over a larger area. This makes it much easier to get full buckets, and provides a constant **back-up** for the operator.

- **Providing a back-up.** A back-up is anything the operator can use to push the load against to get it into the bucket. A back-up is an important element in any loading operation, and also an important consideration when choosing the stockpiling area. Back-ups will be discussed in more detail later in this chapter.

Material Separation

A general rule of thumb to follow when removing material is to keep dirt separate from concrete or other demolition materials at all times. This is especially true in cases where the concrete is to be removed but the dirt remains on the job site for back-filling or some other future use.

One example would be a job where concrete is to be removed from an area especially saw-cut for a trench, which is then dug for a sewer or utility line to be laid. In this situation the best approach would be to pile the broken pavement on one side of the trench, and the dirt on the opposite side. Or, the concrete could be removed first and stockpiled to one side before digging.

There are a number of other instances where the operator will be asked to remove a layer of paving material and then dig an excavation of some type. Unless otherwise requested, **always** keep dirt separate from other materials.

We will now examine separately some demolition and removal techniques that are used with the backhoe and the loader.

Backhoe Techniques

Removal of Concrete

When the backhoe is used to remove concrete, the first thing the operator looks for is an opening or some other space where the bucket teeth can be worked underneath the concrete, such as the edge of a sidewalk or driveway. On many jobs, however, there is no access to an edge, so the operator must create his own starting point by breaking a hole through the concrete where the bucket can gain hold. This is usually easy where the material is three inches thick or less (Fig. 13-2); when it is thicker than three inches, however, breaking with the backhoe could damage the teeth or other machine parts, and is **not recommended.** When it becomes necessary to break thick materials, other equipment should be used, such as a pneumatic pavement breaker, stomper or hydro-hammer (Fig. 13-3), hoe-ram or another tool designed specifically for breaking concrete.

13-2. Using the breaking technique illustrated in Fig. 13-7, the bucket teeth may be used to break through materials up to three inches thick.

In places where the operator can gain access to an edge, however, concrete as thick as six to eight inches can be removed with the backhoe, depending on whether reinforcing steel is contained within the slab. By using the prying method, the material can be lifted and broken at the same time, and then positioned into the bucket for stockpiling or immediate loading.

Prying method. As its name indicates, the prying method is used to loosen, lift and break concrete slabs, curbs, footings, etc. by working the bucket teeth underneath the concrete and then prying it up using the TLB's tremendous hydraulic power.

First, work the teeth under the concrete as far as possible. Once the teeth are in place, the operator exerts downward pressure on the boom and curls the bucket up, while the crowd is extended slightly. Instead of simply trying to lift the concrete, the bucket is used as a fulcrum.

13-3. When thicker material is encountered, access to an edge must usually be created with other types of equipment specially designed for concrete breaking.

The force used to lift the concrete is dissipated into the ground, not to the machine (Fig. 13-4).

The amount of breakout force created by this method is tremendous. It is generated by combining the forces of three of the backhoe's four main hydraulic cylinders (boom, bucket and crowd), plus the weight of the machine. Once the pieces are lifted, they are either loaded onto a truck or stockpiled. If they are to be stockpiled, they should be broken into pieces of the desired size.

Caution is advised when using the prying method. As noted above, the power generated in the prying action is not transferred into the machine, but directly into the ground, **through the bucket.** Because of the tremendous forces generated, and the generally abusive nature of concrete breaking and removal work, using a **heavy-duty bucket** is recommended. Light-duty, special purpose buckets, or buckets which have worn thin due to age, are not recommended. Severe bucket damage could result.

Concrete Breaking

In order to complete common concrete breaking and removal jobs effectively, the sections

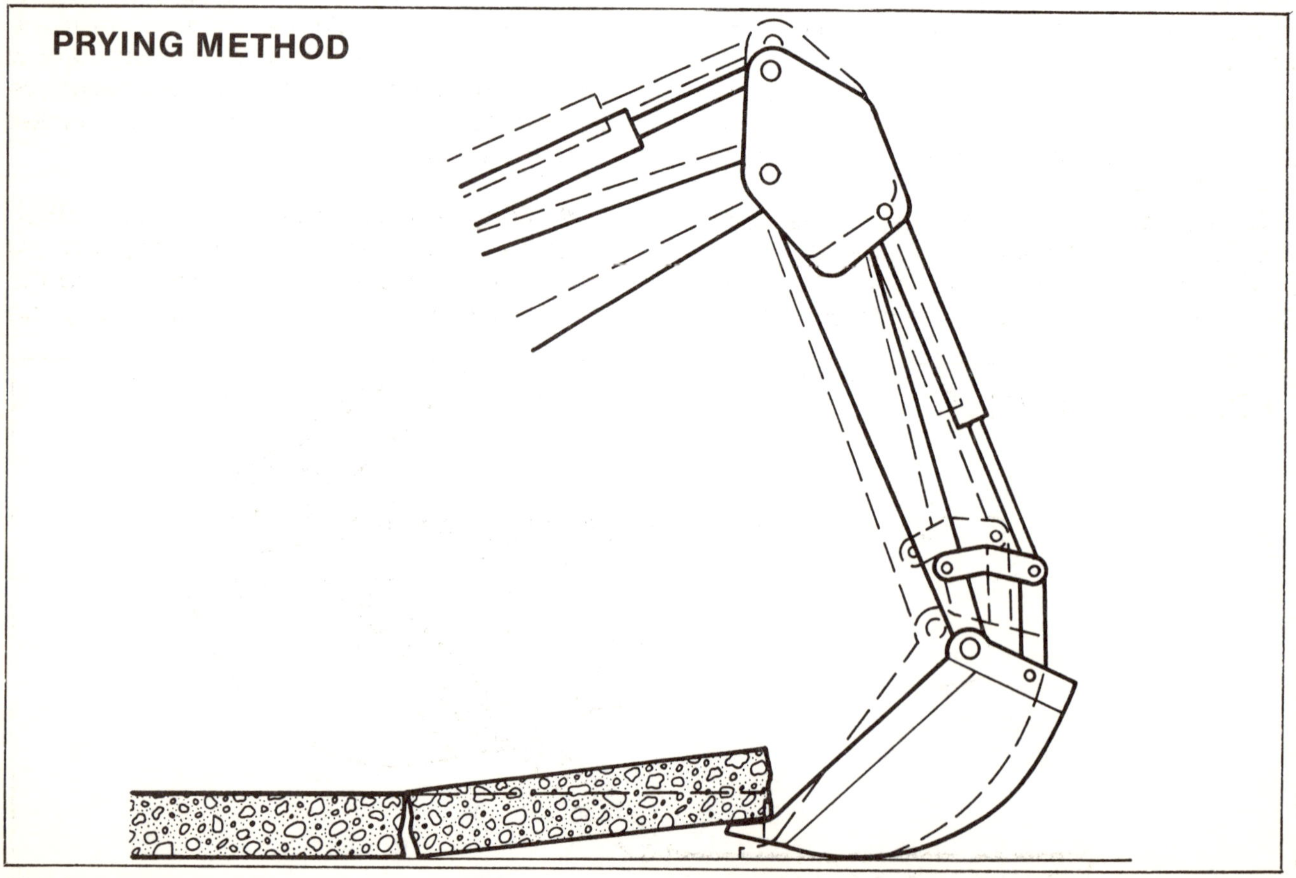

13-4. Once the bucket teeth gain access underneath the concrete the prying method may be used to break and raise material, as illustrated here.

and pieces must be sized properly for eventual loading onto trucks. If the backhoe is to be used, pieces should only be as large as can be safely handled. By the time the loader bucket is used to actually load the material on the truck, pieces should be as small as possible to facilitate the process. As a rule of thumb, concrete pieces should be broken down to the point where they are small enough to load by hand. By breaking the material to this size, the operator realizes two goals: the bucket will load easily, and any material remaining behind can then be loaded by hand.

Recall that handling material as few times as possible during removal is an important goal. However, it should be noted that on more than **half** of the breaking and removal jobs involving the TLB, the material will usually be handled two or more times. In order to avoid unnecessary delays during the second (and subsequent) phases of material handling and loading, materials **must** be broken to a size that will ensure easy handling by a machine of this size.

It must be pointed out that breaking concrete can be **very abusive** to the machine. The operator must learn through experience what his machine can and cannot do, as well as the methods by which concrete may be broken **without** abusing the TLB. The operator should always try to prevent unnecessary machine abuse. Here are several methods by which concrete can be effectively broken, with minimal risk of damage to the TLB.

Dropping. This simple approach is useful when breaking very large, thick slabs and sections into smaller, more manageable pieces. After using the prying method to loosen concrete, it will frequently break off in large sections—too large for the TLB to handle. By picking up one side or end of such a piece, and then moving the bucket out from under it, it will simply fall and break against the remaining concrete. The point is to utilize the concrete itself to help the machine do the breaking, thereby avoiding machine abuse.

The dropping method is also effective when breaking thinner concrete. Because these sections weigh less than thicker slabs, they can be picked up with the bucket, and positioned directly behind the TLB with the crowd extended outward nearly all the way. Then, starting from a height of about five feet, the boom is then raised and lowered very quickly, rolling the bucket out during the downward motion. This short, quick up-and-down motion, combined with the bucket roll, will drop the piece flatly on the ground. This is one of the simplest, most effective ways to break concrete, but the operator should be careful to use it only in areas where it will be completely safe to do so, and where no damage to other structures or objects will occur.

Roll-out method. This is another good breaking method, because it is easy on the TLB and is done with a motion which is natural to the work cycle. It can be used to break long, narrow pieces by positioning them about two feet above the ground, and then simply rolling the bucket outward (Fig. 13-5).

Striking. This is the least desirable method of breaking concrete and carries the greatest risk of machine damage. It involves using the weight of the backhoe assembly to quickly and

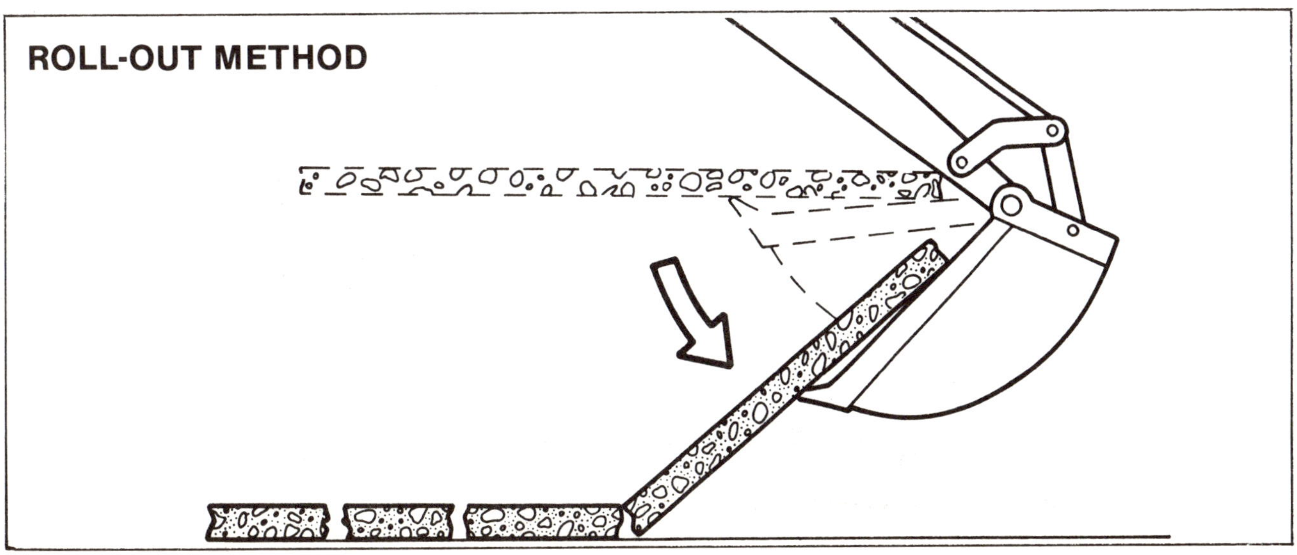

13-5

forcefully strike the concrete with the bucket. The abusive nature of this method can be minimized, however, once the operator gets to know the TLB's limitations, and works within them. The key to using this method safely and effectively is to **correctly position** the pieces prior to striking them, reducing the force necessary for breaking.

Because concrete is a very brittle substance, a light blow will usually break sections up to five inches thick, **if** they are positioned and then struck correctly. To do this, move the piece into a position where it is supported at two points (Fig. 13-6). The piece can then be easily broken at the points indicated by the arrows.

The correct breaking technique involves propelling the boom faster than the hydraulics would normally allow. To accomplish this, position the bucket two to four inches above the precise contact point desired (Fig. 13-7a). Then, raise and lower the bucket as quickly as possible. When done correctly, the maximum distance between the bucket and the concrete at the midpoint between raising and lowering the bucket will be about eight inches (b), and the boom control lever will be returned to the closed position before the bucket contacts the concrete (c). Also, by positioning the crowd at a 90-degree angle to the boom, shocks to the machine will be

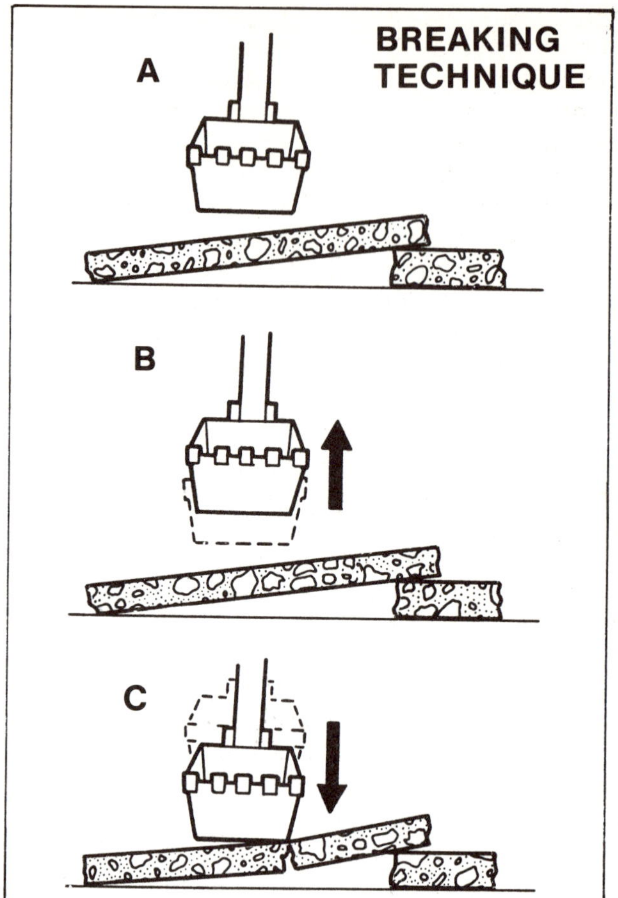

13-7

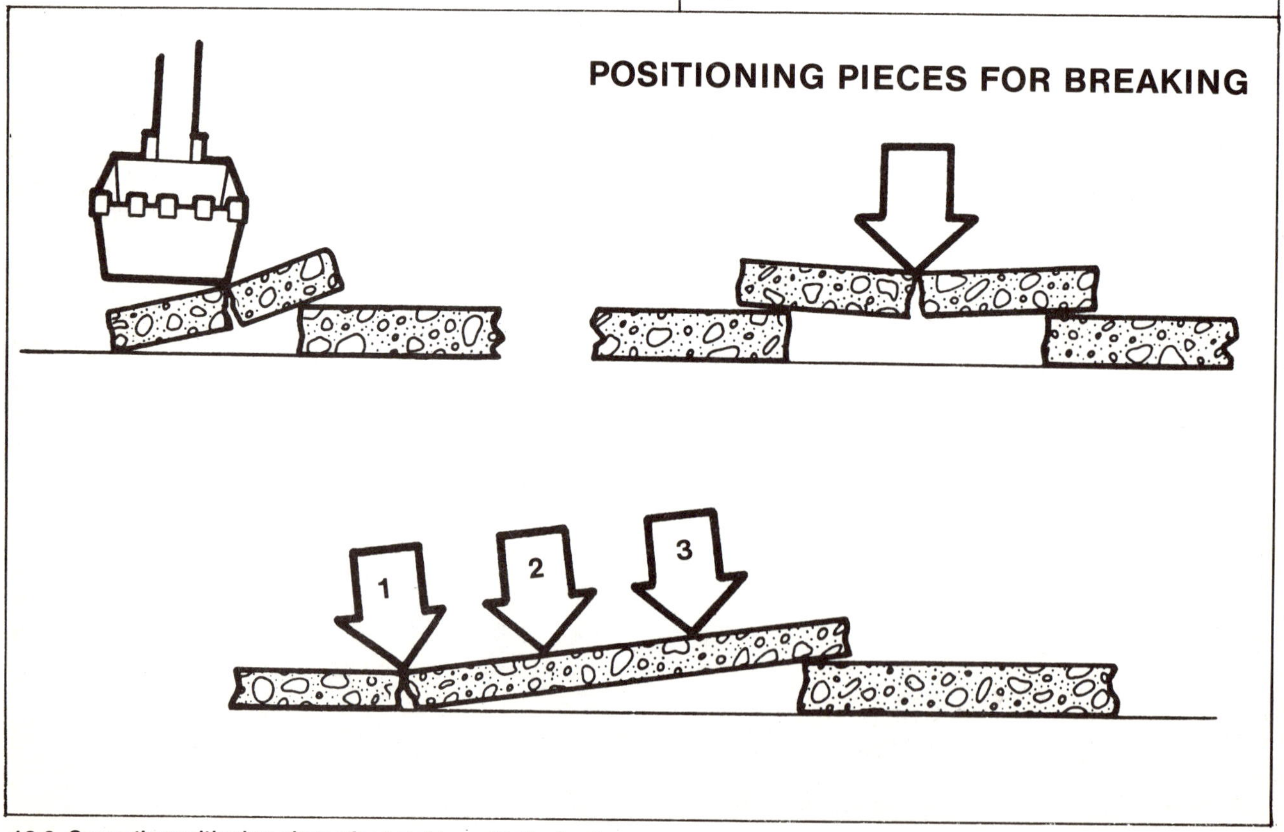

13-6. Correctly positioning pieces for breaking with the bucket.

absorbed by the linkage pins on the bucket, and not by the hydraulic cylinders. Correctly following this procedure will make concrete breaking easier, while preventing pressure surges in the TLB's hydraulic system.

Thickness limitations. It should be clearly understood that all types of equipment have limitations, and that any attempts to exceed those limitations, **especially** when breaking concrete or similar heavy materials, will shorten the life of the machine and could cause immediate, severe structural or hydraulic system damage. The typical breaking limits for the TLB are four to six inches for asphalt, and three to five inches for concrete. These figures are based on correctly positioning the pieces between two points of support, as noted above.

NOTE: All of the breaking methods discussed here do not take into account the possible presence of reinforcing steel in the material. On jobs where steel is a factor to contend with, other methods will be necessary to break materials of this size, or when sizing the material prior to removal.

Alternative breaking methods. There are several types of equipment in common use that are especially designed for breaking concrete, and these may be necessary if the material in question is beyond the limitations of the TLB.

On materials up to one foot in thickness, a "stomper" or hydro-hammer can be very effective. Also, several pneumatic or hydraulically-operated, backhoe-mounted breaking devices are available, and are effective in many situations. In addition, the common pavement breaker or jack-hammer is useful, especially for providing an access hole when using the prying method. This is also an excellent choice when breaking concrete or other materials surrounding areas that are to remain paved, thus preventing damage to these areas.

Loading Techniques

The importance of back-ups. As defined earlier, a back-up can be any object, structure or pile of material that can be used by the operator to push against to facilitate loading. In some applications, the TLB itself can be used as an effective back-up. Providing an adequate back-up is an important part of any operation where repeated filling of the bucket is necessary for transport or loading. This includes operations using either the backhoe or the front loader.

Learning to provide back-ups is especially important for the beginning operator. Because of the unusual shapes and sizes of the pieces to be handled, as well as the weight involved, the beginner will often find loading to be very difficult and awkward at first. Back-ups make the job much easier; using them efficiently makes an operator much more productive as well.

Loading with the Backhoe

In many situations, loading with the backhoe has numerous advantages over alternative loading methods. For example, loading the pieces with the backhoe as they are removed makes one-step removal possible. Loading with the backhoe is also an efficient method of handling large, long or awkward pieces. When handling this type of material, providing adequate back-ups is especially important. This type of loading requires a high degree of operator concentration, smoothness and awareness of safety.

At any time trucks are being loaded, the bucket adjustment must be in the position which provides maximum curl, making it possible to get a good grip on the material. To explain it better, we will separate the loading cycle into two parts: 1) Positioning the bucket, picking up the pieces and gripping the load tightly; and 2) lifting the load, swinging it over the truck and placing it into the truck.

Picking Up the Pieces

The first consideration here is whether the pieces to be picked up are in the **working range** of the backhoe (Fig. 13-8). As loading proceeds, the broken concrete has a natural tendency to be pushed toward the machine. If the pieces are pushed too close to the rear of the machine, they become impossible to pick up. To avoid this, the operator must be aware of this tendency and keep the material away from the area directly behind the machine. Try to keep it pushed back. Notice we said **try**—it seems that no matter how hard one tries, there are always some pieces that flip over and/or are pushed into that area. Once it gets into the area behind the machine, it can't be pushed out directly with the hoe. It can, however, be maneuvered out by carefully using the swing, pushing the pieces sideways, as shown in Fig. 13-9. Repeating this procedure two or three times, back and forth, will easily position the piece back into the working area—within the machine's working range.

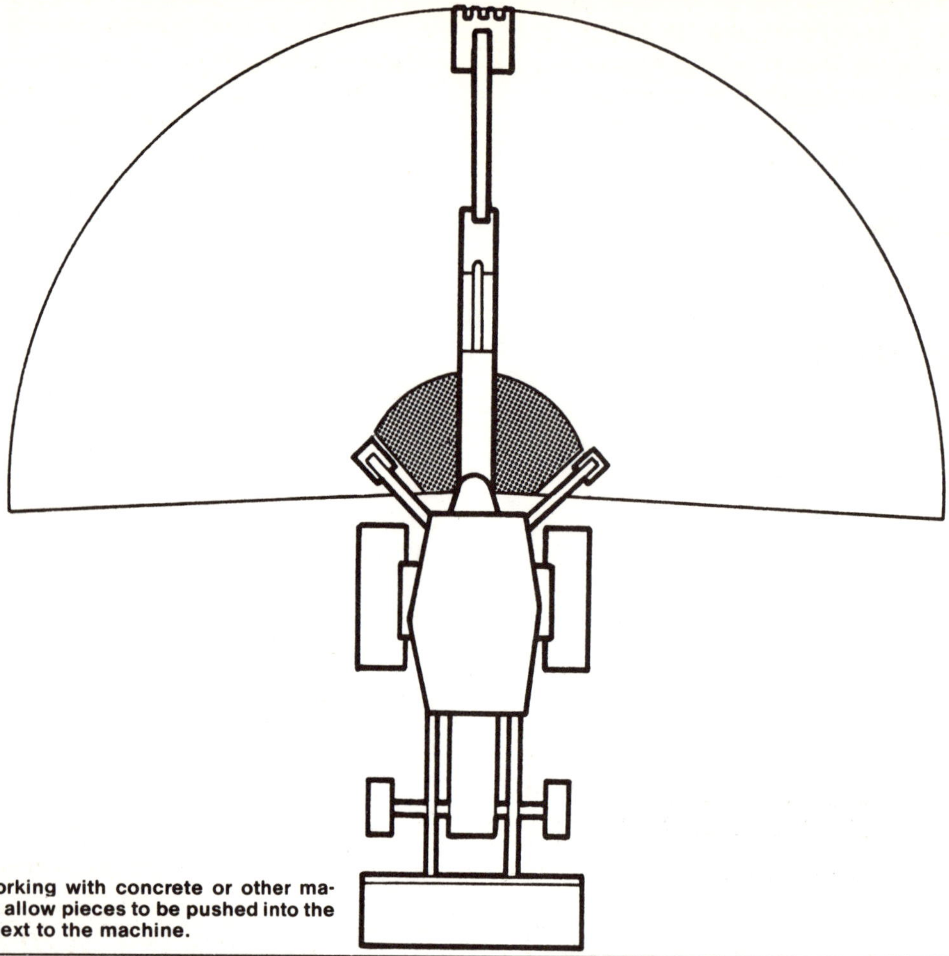

13-8. When working with concrete or other materials, do not allow pieces to be pushed into the shaded area next to the machine.

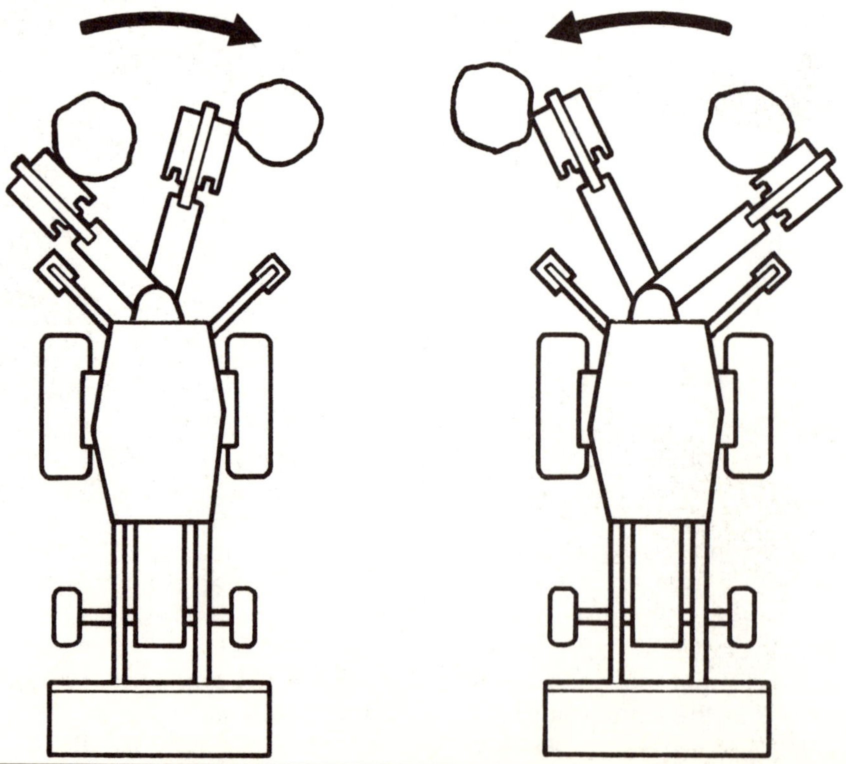

13-9. By pushing objects sideways with the swing they will naturally tend to move away from the machine.

Natural back-ups. When loading material as it is removed, the operator can take advantage of the natural back-up provided by the remaining pavement. This natural back-up provides a perfect means of positioning the bucket under the piece to be removed. Taking advantage of this principle maximizes efficiency whenever picking up pieces as they are broken. (See Fig. 13-10.)

After the piece is broken, the operator must be careful as he positions the bucket under the piece. If the teeth make contact below the piece's center of gravity, it will be lifted over the natural back-up and slide off the bucket and toward the machine. To prevent this, the bucket is lowered and slid under the piece to a position where the teeth are as close to the near edge as possible (Fig. 13-11). As the teeth make contact, the exact same motion as described in the prying method is used to pick up the piece. That is,

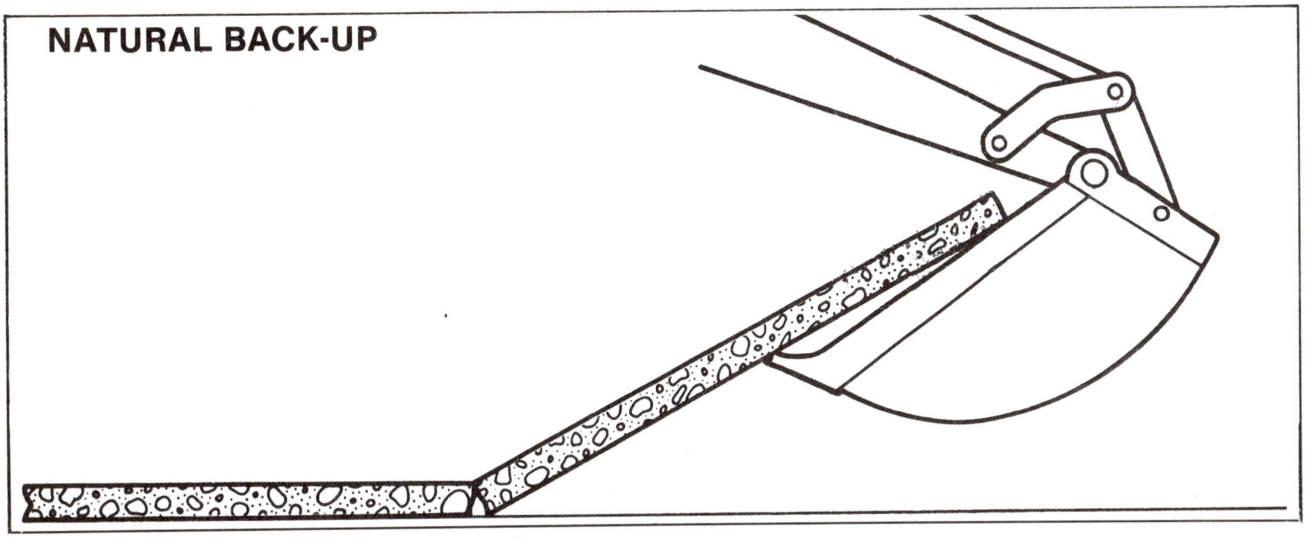

13-10. After using the prying method, the piece will be positioned so that a natural back-up is provided by the remaining material.

13-11

13-12

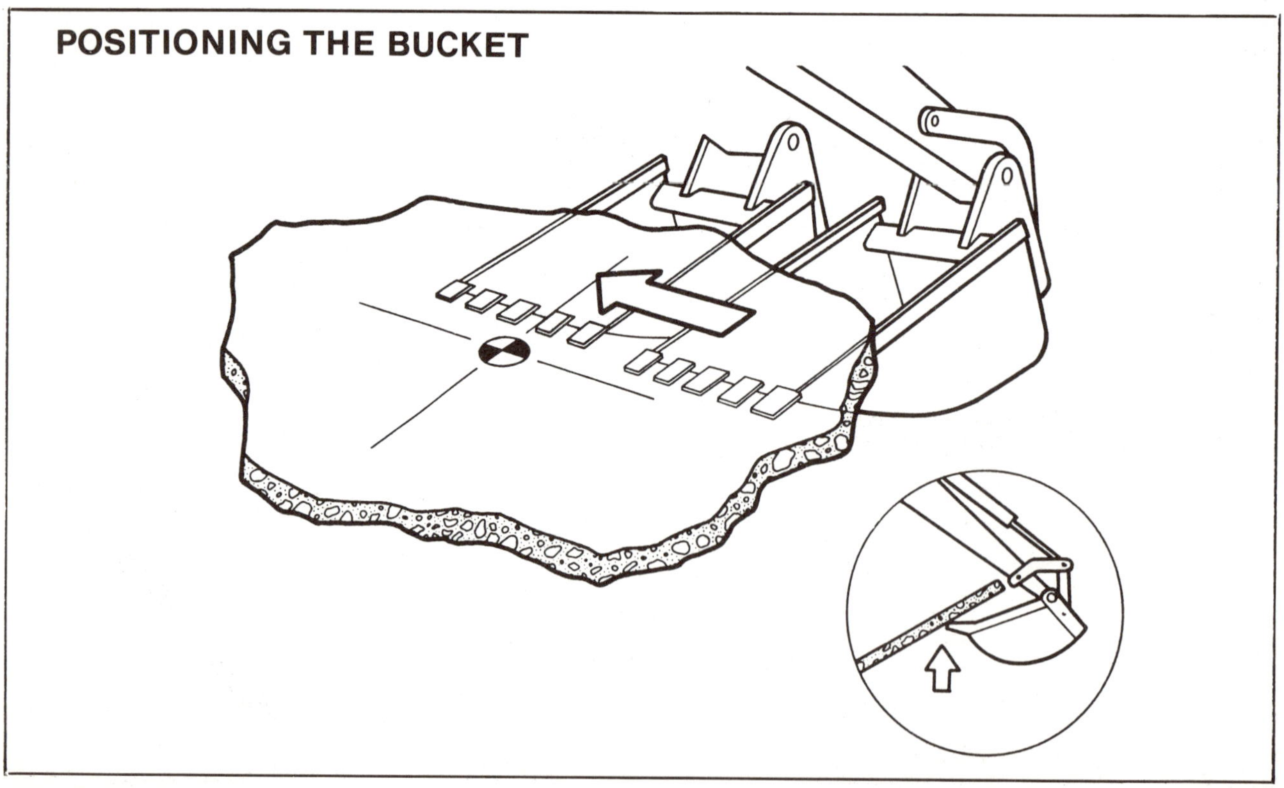

13-13. Illustration shows the correct positioning of the bucket in relation to the piece's center of gravity. Inset shows correct angle at which bucket teeth make contact.

down pressure is applied to the boom, while the bucket is curled up and the crowd extended slightly (Fig. 13-12). After the piece is on the bucket, it is clamped tightly between the bucket and the structure of the crowd.

Positioning the bucket. The correct positioning of the bucket is an important element when picking up material. Often, as the material is lifted and broken, the bucket is not positioned under the piece's center of gravity. To position the bucket in such a way that the piece can be picked up, the bucket must be swung sideways under it. Again, due to the material's heavy weight and rough surface characteristics on the underside, this procedure can be somewhat difficult.

There are several ways to approach this situation. One is to carefully lay the piece down and then pick it up again, with the bucket positioned correctly under the center of gravity. Another method is to reduce the friction between the bucket and the piece, allowing the bucket to slide sideways underneath it. When the weight of the piece is resting on the side (cutting edge) of the bucket, the friction will be too great to allow the necessary bucket movement. To effectively reduce this friction, place the weight of the piece on the bucket teeth, and apply pressure to the swing (Fig. 13-13). When the bucket is in the desired position, the weight should be transferred back to the side of the bucket, to prevent the piece from being lifted over the natural back-up. **Important:** When the teeth contact the piece, the contact point must be above the piece's center of gravity. Contact below that point may lift the piece over the back-up.

Positioning the piece. This is another method that can be used effectively on pieces that are slightly rounded on the bottom. This enables the piece to roll, slightly re-positioning the bucket directly under the piece (Fig. 13-14).

Stacking method. This method works very well when long, narrow sections are to be removed. This is often encountered where pavement has been saw-cut to allow the excavation of a trench. When the material is less than four inches thick, it can usually be stacked up to three layers and easily loaded onto the bucket. As the operator becomes experienced, this method becomes very efficient for removing pavement from a trenching area (Fig. 13-15). **Note:** When pavement is to be cut for a trench, the width of the cut must be wider than the desired trench. For example, if a trench is to be twenty-four inches

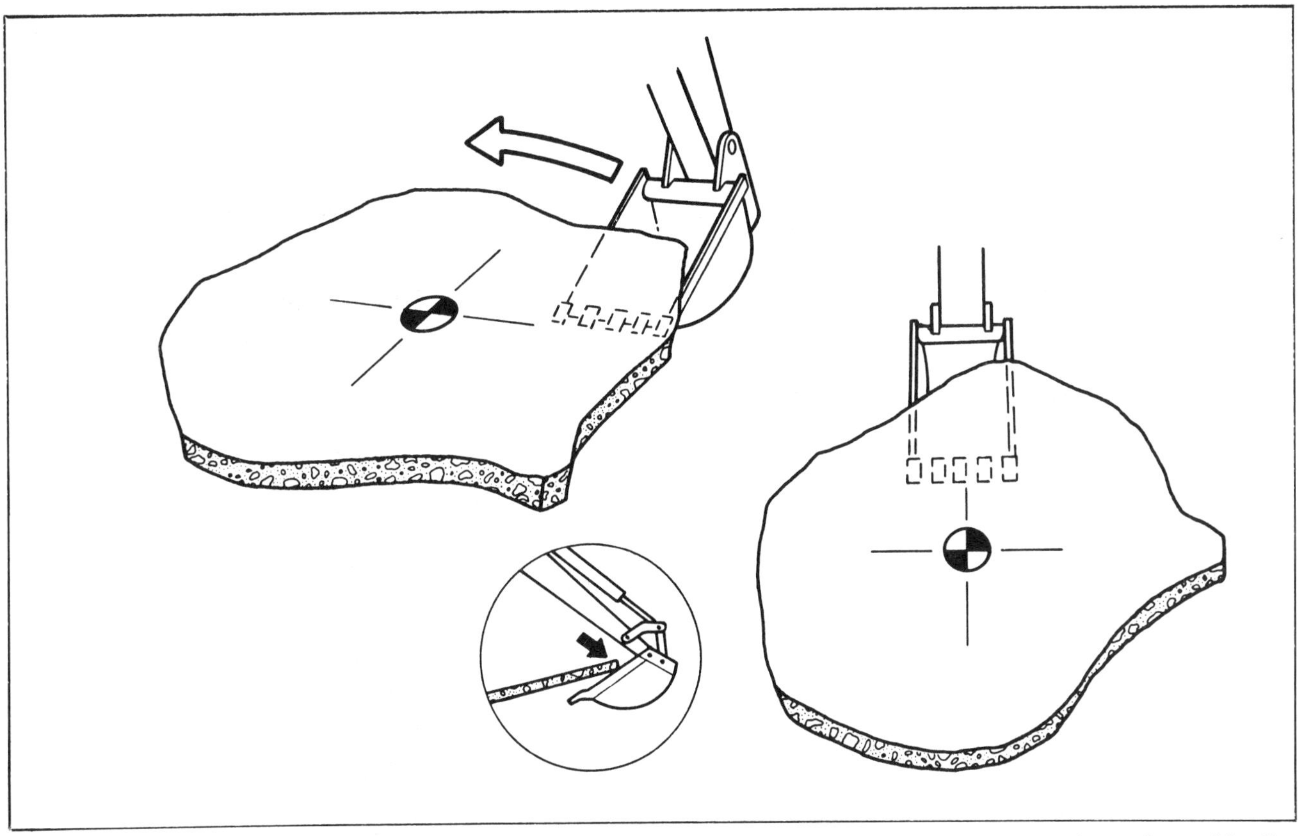

13-14. POSITIONING THE PIECE: When the bucket first contacts the piece incorrectly the swing can sometimes be used to position the piece correctly on the bucket.

wide, the pavement should be cut at least twenty-six inches. If the cut is made exactly to the width of the trench, the bucket simply won't fit into the opening. This in turn means using a smaller bucket, and since it holds less volume, digging will take longer.

Handling thicker concrete. Up to this point, we have referred mostly to handling of slabs four inches thick and less. Thicker material can also be handled effectively with the backhoe (Fig. 13-16). Because thick materials tend to break off in larger pieces, loading is actually easier because the operator will only handle one piece at a time. (It is much easier to handle one large piece than many little ones.) Again, the key to picking up large pieces is to provide a good back-up. When loading large pieces of material, one at a time, the TLB's stabilizers can be used as a very effective back-up where no other object is available.

Using the stabilizer as a back-up. This procedure can be used in many situations. Because of possible damage to the hydraulic ram, however, it is best to find an alternate back-up.

When no other back-up is available, use the utmost care to prevent damage to the ram. To prevent concrete or other heavy, dangerous material from contacting the ram, the material must not be allowed to roll, or slide onto it. This rolling or sliding can be prevented by keeping the ram at least as high as the material which is being loaded into the bucket.

As seen in Fig. 13-17, this maneuver is somewhat difficult and is not recommended for beginning operators. Because of its difficulty and the potential for damage to the TLB, it is easy to understand why this method **should not** be used as standard operating procedure, but as a special, last-resort method. The real beauty of

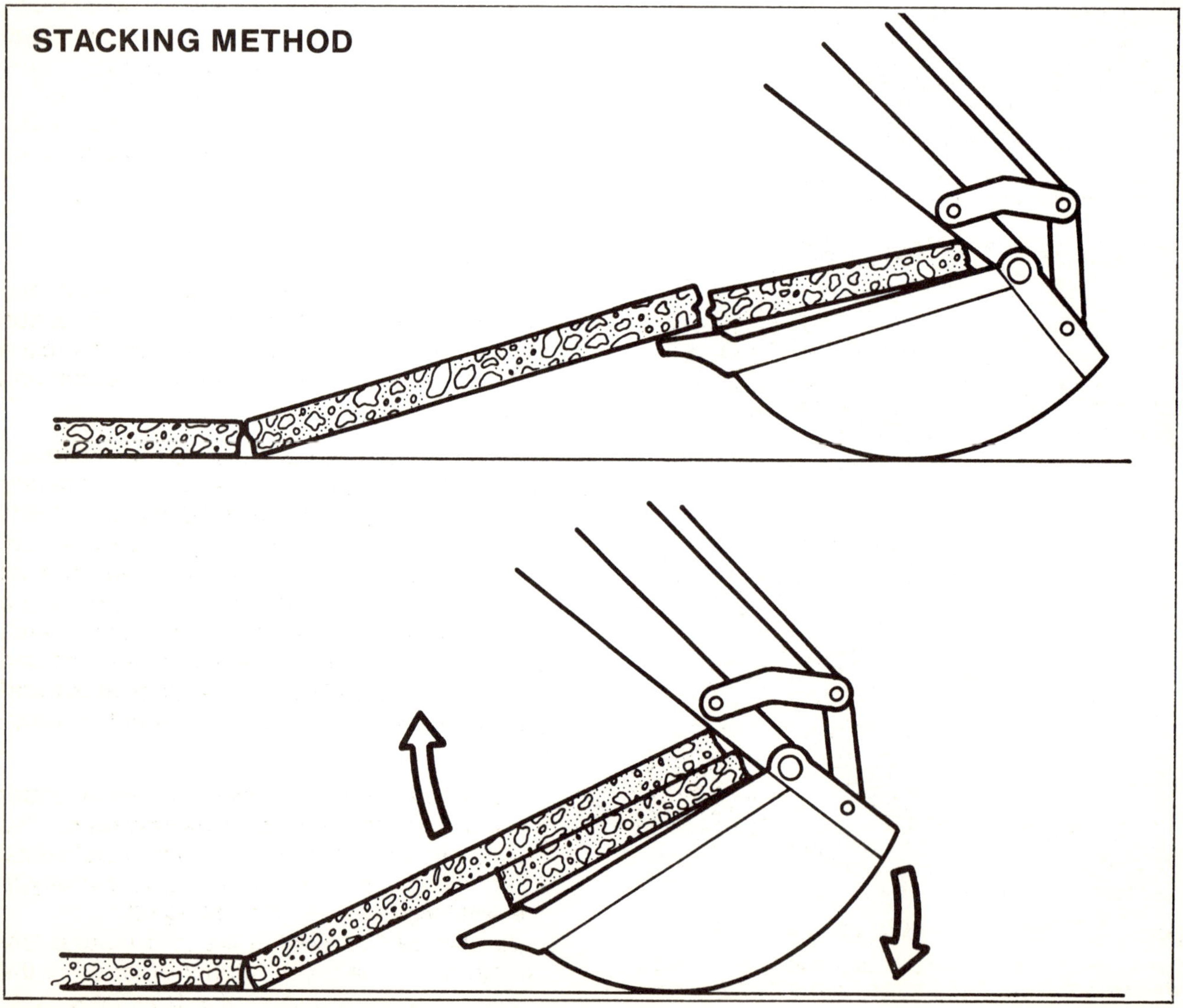

13-15. Long pieces can be broken and stacked for loading in a single motion.

13-16. Thick pieces can also be handled well by loading them in the bucket—one at a time—as shown here.

13-17. USING THE STABILIZER AS A BACK-UP: Pieces can often be pushed onto the bucket by using the stabilizer (beware of damage to the ram).

this technique is realized when it becomes necessary to load large or odd-shaped materials, such as curb and gutter sections, poles, railroad ties, stumps, I-beams, large slabs, etc. Its other obvious advantage is that it is always available, and can be used to eliminate many problem situations.

When attempting this procedure, especially with long pieces of material, it is essential to position the material (or the TLB) so that its center of gravity makes contact **at** the stabilizer. This will ensure the piece is balanced when it is loaded into the bucket. With this in mind, it is easy to see why, when loading a number of long pieces, it is important to arrange them in an orderly manner on the ground, with the center of gravity of each piece located at the same place (Fig. 13-18). Also see **No Back-up,** Page 170.

Now that we have discussed the many aspects of getting the load into the bucket, let's examine the second part of the loading process, which simply consists of lifting and swinging the load over the truck and positioning the material in the truck bed in an orderly fashion. This can be done either with the backhoe or the loader bucket, and we will consider each separately.

Lifting and Swinging

We have seen many ways to efficiently manipulate material into the bucket. The reason there are so many methods is because there are so many different shapes and sizes to contend with.

After the load has been picked up, it is best to clamp it firmly in place to prevent it from falling out. However, clamping the load in the bucket is not always possible. Dropping a load of concrete from ten feet in the air unintentionally is not only inefficient, it is also dangerous, posing a hazard to the machine, the truck, the work area and people in the area. The lifting and swinging of the load must **always** be **very smooth and gentle,** to reduce the risk of accidentally dropping it.

So, when lifting and swinging, the operator must first and foremost **control the load.** This means he must move very smoothly at all times; sometimes, he must also proceed very slowly, to prevent loose pieces from falling off.

As the load is lifted, there is a natural tendency for the pieces to fall out of, or off of the bucket. During the course of a day's work, there

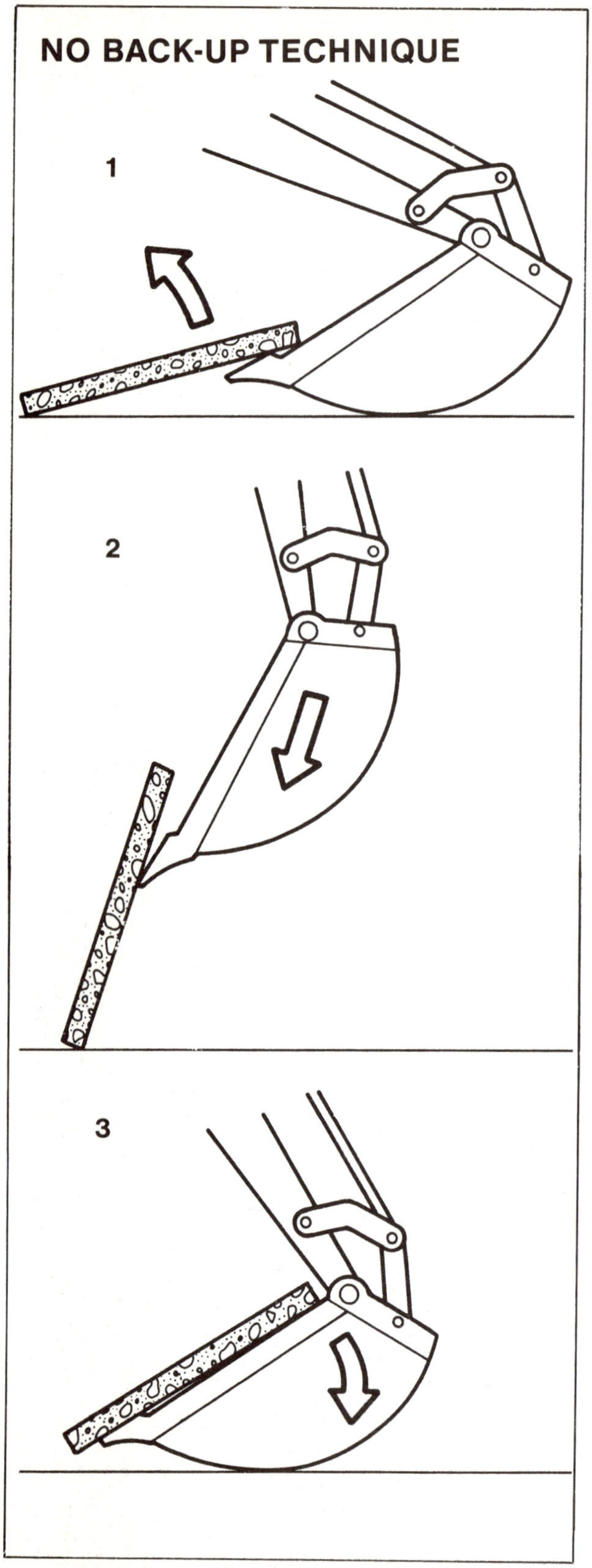

13-18. NO BACK-UP TECHNIQUE: When no back-up is available, this three-step procedure is recommended: 1) The piece is lifted until near vertical. 2) The bucket slides carefully down behind the piece. By rolling the bucket down and back (3), the piece will be secured on the bucket.

will be many situations when a load may contain pieces that are not securely held in place (Fig. 13-19). When it appears that all or part of a load may fall as it is swung over the truck, the operator may want to "test" the load to see how securely it is held in place. The purpose is to make sure the load will be safe when swinging it to the side. If it isn't, falling pieces could damage the truck, the TLB's stabilizer arm, or, if the load is very high, it could even fall onto the cab area of the machine. When conducting this test, the load is positioned straight behind the machine, in a manner that would be safe if pieces do fall.

Position the load directly behind the TLB over an area where falling concrete would not present a problem. Clamp it as tightly as possible with the bucket. Then, a very light and quick movement of the crowd will shake the load so that any pieces not securely held will move or fall. The bucket is then re-tightened, and the load should be safe to swing, as the remaining pieces are securely gripped by the machine (Fig. 13-20).

As the load swings over the truck and clears the side, it should be lowered with the crowd and placed carefully into the truck (not dropped) to reduce the risk of damage. As the pieces are placed into the truck, the operator should pay attention to the load's **weight distribution** in the

13-19. When loading loose, awkwardly-shaped materials as shown here, every move must be made slowly and smoothly to prevent material from falling from the bucket.

bed. The weight of the load should be distributed evenly to both sides.

Another important point is to place the materials in such a way that they won't bind the load when it is dumped later on. After large slabs are placed in the bed, they should then be carefully broken by hand to make sure the load will dump easily.

One way to prevent a load from binding is not to pile it too high, never higher than the tailgate hinge pin. If piled too high, the load will slide back and bind as the bed is raised, making it impossible to lower the bed again until the material if off-loaded by hand.

Loading with the Loader

When loading out concrete with the loader bucket, the same rules apply as when loading any other type of material. Loading concrete requires a little extra planning and preparation, however, if efficient removal and minimal problems are the operator's goals. The two main aspects to be considered are 1) the height of the stockpile, and 2) the size of the pieces.

As stressed earlier, keeping the stockpile as low as possible is the best approach (see Figs. 13-21, 22, 23). The lower the pile, the easier it will be to fill the bucket. In general, material should never be piled higher than is absolutely necessary.

The size of the pieces is also very important. The smaller they are, the easier it will be to load them into the bucket. If larger pieces are stockpiled without being broken first, they can be very difficult to load. In contrast, when crushed or broken into pieces less than two inches, the bucket can be filled quickly and easily. Of course, pieces will rarely be broken up that small on most jobs, but the principle remains: **small pieces are easier to load than large ones.** The only exception is when large pieces are loaded one at a time.

Large pieces. Loading large chunks of material with the loader can be a very dangerous operation. There is no way to "grip" them with the loader, as the backhoe can do.

Due to the weight of the large pieces, a smooth and nearly level loading area is a necessity. Because the TLB is designed to lift only as much as it can safely handle, it is fairly easy to avoid trouble as far as over-loading the machine is concerned. However, this applies on **level ground only.** Loading heavy material on any sloped surface is extremely dangerous, and should not be attempted.

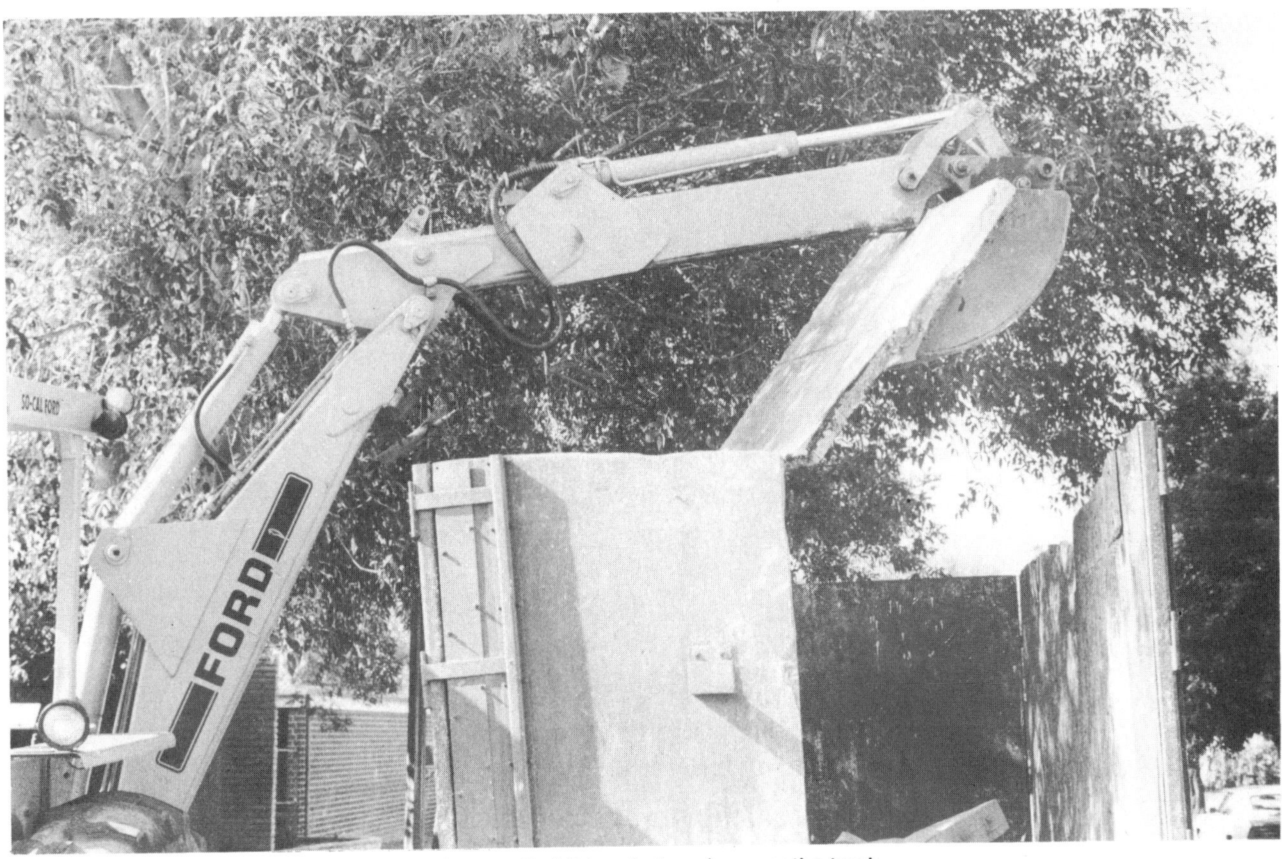

13-20. By testing the load, the operator can be sure that it is safe to swing over the truck.

13-21. Even when a pile like the one shown here is only as high as the bucket, resistance is often so great that getting a full bucket will often be impossible.

13-22. Here the pile was back-dragged and effectively lowered so that the material can be pushed easily into the bucket.

13-23. When loading from a properly prepared stockpile, getting full buckets will be consistently easier.

13-24. After the piece has been lifed as shown, the leading edge of the bucket is worked down and under the piece as far as possible.

13-25. If the top of the piece is heavier than the bottom, it will naturally fall onto the bucket—ready to load.

The recommended method for loading large pieces into the front loader bucket is similar to that for loading the backhoe. First, position the piece where a back-up can be used to force the piece into the bucket. Then, lift the piece to about a 45 to 60 degree angle (Fig. 13-24). Carefully work the bucket down and under the piece. As the lower edge of the bucket is positioned as far below the piece's center of gravity as possible, the bucket can be rolled back. If the piece does not remain in the bucket, roll the bucket up and down repeatedly, **working it down and under the piece.**

Once the piece is resting in the bucket (Fig. 13-25) (or in the case of large, flat pieces, lying on the outer bucket surfaces), all movements must be smoothly executed to keep the load from shifting during transport or when raising the loader. Just as when loading is done with the backhoe, the pieces must be placed carefully in the truck bed (not dropped) to prevent damage. Before loading large pieces, a cushioning layer of dirt, or small concrete pieces, should be placed in the bed to further protect it. As before, load distribution is a factor—distribute the material evenly in the truck, during all loading operations.

CAUTION: Machines which are not equipped with **self-leveling buckets** can be extremely dangerous when loading out concrete. Constant attention to keeping the bucket level is required to prevent pieces from falling onto the cab area of the machine.

13-26. This photo shows a common demolition job site shortly after work has begun.

13-27. Here the bulk of the material has been removed.

13-28. Job site is shown following clean-up.

GLOSSARY OF TERMS

BACK-UP: Any object or structure the operator may use to push against, moving material into either the backhoe or loader bucket.

BOOM: The main lifting structure of the hoe assembly.

BREAKOUT FORCE: The force created by the curling action of the loader bucket.

CAVE-IN: The sudden collapse of the excavation wall.

CENTER OF GRAVITY: The point within the machine around which its weight is evenly distributed.

COMPACTION: The compression of soil into a dense state through the application of great weight, usually that of the machine.

CONDUIT: Plastic or steel piping through which wires pass underground.

CONTROL: The ability to effectively direct the TLB so that it performs exactly as the operator desires.

CROWD: The secondary structure of the hoe assembly, to which the backhoe bucket is attached.

DEMOLITION: Breaking and removal of concrete, asphalt or other heavy building materials.

DIGGING CYCLE: The series of motions used by the operator during excavating, from filling the bucket, to swinging and dumping the bucket, to returning to the original digging position.

DISPERSION: To distribute or spread material in an orderly manner aiding in job completion.

DITCH: See ***trench.***

ELEVATION: An established point above sea level or other measured point of reference.

EXCAVATION: Any digging operation involving the removal of earth; also, the space created by digging.

FEATHERING: Process of manipulating the hydraulic control valves on the TLB so that a smooth, flowing action is created.

FLOAT: Position in the loader control valve which allows the loader arms to raise and lower freely.

FOOTINGS: The concrete foundation of a building or other structure, requiring excavations of a precise nature.

GRADE: A) To cut, fill and smooth a given area. B) See ***elevation.***

HOSES: Flexible rubber/steel/fabric pipes which carry fluid through the TLB's hydraulic system.

JETTING: Process of injecting great amounts of water through a hose to speed the compaction process.

JOYSTICK: Control lever which combines two functions, and which is manipulated in four directions instead of two.

LAYOUT: A job plan, usually laid on the ground with lime or other visible material, to indicate the direction and/or trench width for the TLB to follow.

LEVEL: A perfectly flat or even surface area, conforming to the surface of still water.

MONITOR: To periodically check the status of; watch over.

OVERDIGGING: Digging too deeply or farther than intended.

PIER: A large concrete platform which supports the weight of a building or other structure.

PIVOT METHOD: The technique of maneuvering the TLB using the loader bucket as a pivot point.

PLUMB: Perpendicular to level.

RESTRICTIONS: Obstacles or other hazards which prevent free movement or operation of the TLB.

ROLL-OVER: Accidentally causing the TLB to tip over violently.

SET-UP: Exact positioning of the TLB that will produce the desired excavation results.

SHANK: Permanent mounting to which bucket teeth are attached.

SHORING: A system of wood or metal restraints which, when placed inside an excavation, hold the walls in place and prevent cave-ins.

SHUTTLE: The repeated movement back and forth of the TLB across the job site in a predetermined route.

SLOPING: The digging of excavation walls in such a way, and at such an angle, that the risk of cave-ins is eliminated; an alternative to ***shoring.***

SPOIL: Material removed from an excavation.

STICK: See ***crowd.***

STOCKPILE: The accumulation of spoil or other materials on the job site, in areas which are accessible to the TLB, trucks, etc.

SWING: The hoe function which permits side movement.

TEETH: The hardened steel cutting edges which are attached to the backhoe bucket.

TOOTH-GRAB METHOD: Technique by which the operator may use the bucket teeth to hook or grab objects such as small trees, pipes, reinforcing rods, etc.

TORQUE: The force which produces rotation; in the TLB, the reaction to the force takes the form either of lifting force or down pressure on the bucket or front tires.

TOTAL OPERATING WEIGHT: The weight of the TLB itself plus the load it is transporting.

TRACTION: The amount of friction between the tires and the surface traveled on.

TRENCH: A particular type of excavation, usually no wider than the bucket being used to dig it.

UTILITIES: The presence on a job site of lines operated by public agencies, including electrical, telephone, gas or water lines.

WEIGHT DISTRIBUTION: The manner in which the total operating weight of the TLB is distributed into the area on which it sits; may change depending on a variety of external factors.

ZERK FITTING: Common fittings located around the TLB, through which a grease gun may be attached for the injection of grease into the operating hinge joints.

Index

Other Practical References

Construction Estimating References

Construction Estimating Reference Data
Collected in this single volume are the building estimator's 300 most useful estimating reference tables. Labor requirements for nearly every type of construction are included: sitework, concrete work, masonry, steel, carpentry, thermal & moisture protection, doors and windows, finishes, mechanical and electrical. Each section explains in detail the work being estimated and gives the appropriate crew size and equipment needed. Many pages of illustrations, estimating pointers and explanations of the work being estimated are also included. This is an essential reference for every professional construction estimator. **368 pages, 11 x 8½, $18.00**

Building Cost Manual
Square foot costs for residential, commercial, industrial, and farm buildings. In a few minutes you work up a reliable budget estimate based on the actual materials and design features, area, shape, wall height, number of floors and support requirements. Most important, you include all the important variables that can make any building unique from a cost standpoint. **240 pages, 8½ x 11, $12.00. Revised annually**

Construction Industry Production Manual
Manhour tables developed by professional estimators from hundreds of jobs and all types of construction. Thousands of carefully researched figures, accurate charts and precise tables to give the estimator the information needed to compile an accurate estimate. If you have only one book of labor tables, this is the book to have. **176 pages, 5½ x 8½, $8.00**

Repair and Remodeling Cost Estimator
Cost estimates for repair and remodeling work for all residential and light commercial buildings: the cost of materials, the quantity of material needed, the appropriate work crew, manhours required to complete the job, total cost per installed unit, and your bid price — based on the size and type of job you're bidding. This is reliable, specific information you can use on every repair and remodeling project. **256 pages, 8½ x 11, $16.75. Revised annually.**

National Construction Estimator
Current building costs in dollars and cents for residential, commercial and industrial construction. Prices for every commonly used building material, and the proper labor cost associated with installation of the material. Everything figured out to give you the "in place" cost in seconds. Many time-saving rules of thumb, waste and coverage factors and estimating tables are included. **512 pages, 8½ x 11, $16.00. Revised annually.**

Cost Records for Construction Estimating
How to organize and use cost information from jobs just completed to make more accurate estimates in the future. Explains how to keep the cost records you need to reflect the time spent on each part of the job. Shows the best way to track costs for sitework, footing, foundations, framing, interior finish, siding and trim, masonry, and subcontract expense. Provides sample forms. **208 pages, 8½ x 11, $15.75**

Electrical References

Electrical Construction Estimator
If you estimate electrical jobs, this is your guide to current material costs, reliable manhour estimates per unit, and the total installed cost for all common electrical work: conduit, wire, boxes, fixtures, switches, outlets, loadcenters, panelboards, raceway, duct, signal systems, and more. Explains what every estimator should know before estimating each part of an electrical system. **400 pages, 8½ x 11, $25.00**

Residential Electrical Design
Explains what every builder needs to know about designing electrical systems for residential construction. Shows how to draw up an electrical plan from the blueprints, including the service entrance, grounding, lighting requirements for kitchen, bedroom and bath and how to lay them out. Explains how to plan electrical heating systems and what equipment you'll need, how to plan outdoor lighting, and much more. If you are a builder who ever has to plan an electrical system, you should have this book. **194 pages, 8½ x 11, $11.50**

Estimating Electrical Construction
A practical approach to estimating materials and labor for residential and commercial electrical construction. Written by the A.S.P.E. National Estimator of the Year, it explains how to use labor units, the plan take-off and the bid summary to establish an accurate estimate. Covers dealing with suppliers, pricing sheets, and how to modify labor units. Provides extensive labor unit tables, and blank forms for use in estimating your next electrical job. **272 pages, 8½ x 11, $19.00**

Manual of Electrical Contracting
From the tools you need for installing electrical work in new construction and remodeling work to developing the finances you need to run your business. Shows how to draw up an electrical plan and design the correct lighting within the budget you have to work with. How to calculate service and feeder loads, service entrance capacity, demand factors, and install wiring in residential, commercial, and agricultural buildings. Covers how to make sure your business will succeed before you start it, and how to keep it running profitably. **224 pages, 8½ x 11, $17.00**

Handbook of Modern Electrical Wiring
The journeyman electrician's guide to planning the job and doing professional quality work on any residential or light commercial project. Explains how to use the code, how to calculate loads and size conductors and conduit, the right way to lay out the job, how to wire branch and feeder circuits, selecting the right service equipment, and much more. **204 pages, 5½ x 8½, $14.75**

Electrical Blueprint Reading
Shows how to read and interpret electrical drawings, wiring diagrams and specifications for construction of electrical systems in buildings. Shows how a typical lighting plan and power layout would appear on the plans and explains what the contractor would do to execute this plan. Describes how to use a panelboard or heating schedule and includes typical electrical specifications. **128 pages, 8½ x 11, $8.50**

Remodeling References

Manual of Professional Remodeling
This is the practical manual of professional remodeling written by an experienced and successful remodeling contractor. Shows how to evaluate a job and avoid 30-minute jobs that take all day, what to fix and what to leave alone, and what to watch for in dealing with subcontractors. Includes chapters on calculating space requirements, repairing structural defects, remodeling kitchens, baths, walls and ceilings, doors and windows, floors, roofs, installing fireplaces and chimneys (including built-ins), skylights, and exterior siding. Includes blank forms, checklists, sample contracts, and proposals you can copy and use. **400 pages, 8½ x 11, $18.75**

How to Sell Remodeling
Proven, effective sales methods for repair and remodeling contractors: finding qualified leads, making the sales call, identifying what your prospects really need, pricing the job, arranging financing, and closing the sale. Explains how to organize and staff a sales team, how to bring in the work to keep your crews busy and your business growing, and much more. Includes blank forms, tables, and charts. **240 pages, 8½ x 11, $17.50**

Remodelers Handbook
The complete manual of home improvement contracting: Planning the job, estimating costs, doing the work, running your company and making profits. Pages of sample forms, contracts, documents, clear illustrations and examples. Chapters on evaluating the work, rehabilitation, kitchens, bathrooms, adding living area, re-flooring, re-siding, re-roofing, replacing windows and doors, installing new wall and ceiling cover, re-painting, upgrading insulation, combating moisture damage, estimating, selling your services, and bookkeeping for remodelers. **416 pages, 8½ x 11, $18.50**

Paint Contractor's Manual
How to start and run a profitable paint contracting company: getting set up and organized to handle volume work, avoiding the mistakes most painters make, getting top production from your crews and the most value from your advertising dollar. Shows how to estimate all prep and painting. Loaded with manhour estimates, sample forms, contracts, charts, tables and examples you can use. **224 pages, 8½ x 11, $19.25**

Plumbing & Piping References

Plumbers Handbook Revised
This new edition shows what will and what will not pass inspection in drainage, vent, and waste piping, septic tanks, water supply, fire protection, and gas piping systems. All tables, standards, and specifications are completely up-to-date with recent changes in the plumbing code. Covers common layouts for residential work, how to size piping, selecting and hanging fixtures, practical recommendations and trade tips. This book is the approved reference for the plumbing contractors exam in many states. **240 pages, 8½ x 11, $16.75**

Estimating Plumbing Costs
Offers a basic procedure for estimating materials, labor, and direct and indirect costs for residential and commercial plumbing jobs. Explains how to interpret and understand plot plans, design drainage, waste, and vent systems, meet code requirements, and make an accurate take-off for materials and labor. Includes sample cost sheets, manhour production tables, complete illustrations, and all the practical information you need to accurately estimate plumbing costs. **224 pages, 8½ x 11, $17.25**

Plumber's Exam Preparation Guide
Lists questions like those asked on most plumber's exams. Gives the correct answer to each question, under both the Uniform Plumbing Code and the Standard Plumbing Code — and explains why that answer is correct. Includes questions on system design and layout where a plan drawing is required. Covers plumbing systems (both standard and specialized), gas systems, plumbing isometrics, piping diagrams, and as much plumber's math as the examination requires. Suggests the best ways to prepare for the exam, how and what to study and describes what you can expect on exam day. At the end of the book is a complete sample exam that can predict how you'll do on the real tests. **320 pages, 8½ x 11, $21.00**

Basic Plumbing with Illustrations
The journeyman's and apprentice's guide to installing plumbing, piping and fixtures in residential and light commercial buildings: how to select the right materials, lay out the job and do professional quality plumbing work. Explains the use of essential tools and materials, how to make repairs, maintain plumbing systems, install fixtures and add to existing systems. **320 pages, 8½ x 11, $17.50**

Heavy Construction References

Construction Superintending
Explains what the "super" should do during every job phase from taking bids to project completion on both heavy and light construction: excavation, foundations, pilings, steelwork, concrete and masonry, carpentry, plumbing, and electrical. Explains scheduling, preparing estimates, record keeping, dealing with subcontractors, and change orders. Includes the charts, forms, and established guidelines every superintendent needs. **240 pages, 8½ x 11, $22.00**

Building Layout
Shows how to use a transit to locate the building on the lot correctly, plan proper grades with minimum excavation, find utility lines and easements, establish correct elevations, lay out accurate foundations and set correct floor heights. Explains planning sewer connections, leveling a foundation out of level, using a story pole and batterboards, working on steep sites, and minimizing excavation costs. **240 pages, 5½ x 8½, $11.75**

Process & Industrial Pipe Estimating
A clear, concise guide to estimating costs of fabricating and installing underground and above ground piping. Includes types of pipes and fittings, valves, filters, strainers, and other in-line equipment commonly specified, and their installation methods. Shows how a take-off is consolidated on the estimate form and the bid estimate derived using the complete set of manhour tables provided in this complete manual of pipe estimating. **240 pages, 8½ x 11, $18.25**

Excavation and Grading Handbook
The foreman's and superintendent's guide to highway, subdivision and pipeline jobs: how to read plans and survey stake markings, set grade, excavate, compact, pave and lay pipe on nearly any job. Includes hundreds of practical tips, pictures, diagrams and tables that even experienced "pros" should have. **320 pages, 5½ x 8½, $15.25**

Construction References

Wood-Frame House Construction
From the layout of the outer walls, excavation and formwork, to finish carpentry, and painting, every step of construction is covered in detail with clear illustrations and explanations. Everything the builder needs to know about framing, roofing, siding, insulation and vapor barrier, interior finishing, floor coverings, and stairs. . .complete step by step "how to" information on what goes into building a frame house. **240 pages, 8½ x 11, $11.25. Revised edition**

Contractor's Guide to the Building Code
Explains in plain English exactly what the Uniform Building Code requires and shows how to design and construct residential and light commercial buildings that will pass inspection the first time. Suggests how to work with the inspector to minimize construction costs, what common building short cuts are likely to be cited, and where exceptions are granted. **312 pages, 5½ x 8½, $16.25**

Finish Carpentry
The time-saving methods and proven shortcuts you need to do first class finish work on any job: cornices and rakes, gutters and downspouts, wood shingle roofing, asphalt, asbestos and built-up roofing, prefabricated windows, door bucks and frames, door trim, siding, wallboard, lath and plaster, stairs and railings, cabinets, joinery, and wood flooring. **192 pages, 8½ x 11, $10.50**

Home Builders Guide
The "how to" of custom home building explained by a successful professional builder: How to anticipate problems, eliminate bottlenecks, keep the work going smoothly, and end up with a finished home that is a credit to your skill and professionalism. Explains working with subcontractors, lenders, architects, municipal authorities, building inspectors, tradesmen and suppliers, avoiding design problems, getting financing, making sure your building permit is issued promptly, work scheduling and more. **359 pages, 5½ x 8½, $7.00**

Stair Builders Handbook
If you know the floor to floor rise, this handbook will give you everything else: the number and dimension of treads and risers, the total run, the correct well hole opening, the angle of incline, the quantity of materials and settings for your framing square for over 3,500 code approved rise and run combinations—several for every 1/8 inch interval from a 3 foot to a 12 foot floor to floor rise. **416 pages, 8½ x 5½, $12.75**

Roof Framing
Frame any type of roof in common use today, even if you've never framed a roof before. Shows how to use a pocket calculator to figure any common, hip, valley, and jack rafter length in seconds. Over 400 illustrations take you through every measurement and every cut on each type of roof: gable, hip, Dutch, Tudor, gambrel, shed, gazebo and more. **480 pages, 5½ x 8½, $19.50**

Rough Carpentry
All rough carpentry is covered in detail: sills, girders, columns, joists, sheathing, ceiling, roof and wall framing, roof trusses, dormers, bay windows, furring and grounds, stairs and insulation. Many of the 24 chapters explain practical code approved methods for saving lumber and time without sacrificing quality. Chapters on columns, headers, rafters, joists and girders show how to use simple engineering principles to select the right lumber dimension for whatever species and grade you are using. **288 pages, 8½ x 11, $14.50**

Rafter Length Manual
Complete rafter length tables and the "how to" of roof framing. Shows how to use the tables to find the actual length of common, hip, valley and jack rafters. Shows how to measure, mark, cut and erect the rafters, find the drop of the hip, shorten jack rafters, mark the ridge and much more. Has the tables, explanations and illustrations every professional roof framer needs. **369 pages, 8½ x 5½, $10.75**

Reducing Home Building Costs
Explains where significant cost savings are possible and shows how to take advantage of these opportunities. Six chapters show how to reduce foundation, floor, exterior wall, roof, interior and finishing costs. Three chapters show effective ways to avoid problems usually associated with bad weather at the jobsite. Explains how to increase labor productivity. **224 pages, 8½ x 11, $10.25**

Building and Remodeling for Energy Savings
The practical handbook of what you can and should be doing to meet the needs of your energy-conscious clients: How to build energy efficiency into every job, plan the site to save up to 30% on heating and cooling costs, when to use and when to avoid storm sash and thermal glass, how to use windows, skylights and clerestories to light room interiors, what every builder must know about insulation, weatherstripping and condensation. Hundreds of practical tips and pages of reference data are included. **320 pages, 8½ x 11, $15.00**

Spec Builder's Guide
Explains how to plan and build a home, control your construction costs, and then sell the house at a price that earns a decent return on the time and money you've invested. Includes professional tips to ensure success as a spec builder: how government statistics help you judge the housing market, cutting costs at every opportunity without sacrificing quality, and taking advantage of construction cycles. Every chapter includes checklists, diagrams, charts, figures, and estimating tables. **448 pages, 8½ x 11, $24.00**

HVAC Contracting
Your guide to setting up and running a successful HVAC contracting company. Shows how to plan and design all types of systems for maximum efficiency and lowest cost — and explains how to sell your customers on the designs you propose. Describes the right way to use all the instruments, equipment and reference materials essential to HVAC contracting. Includes a full chapter on estimating, bidding, and contract procedure. **208 pages, 8½ x 11, $24.50**

Builder's Guide to Government Loans
This comprehensive guide will help you take advantage of the many government loan programs. Everything is explained in step-by-step instruction and actual sample forms are included: HUD and FHA loan insurance and guarantees for operative builders, interest subsidies, rent supplements, public housing programs, turnkey projects, VA loans, FmHA rural housing programs, and Small Business Administration loans for builders. **416 pages, 5½ x 8½, $13.75**

Roofers Handbook
The journeyman roofer's complete guide to wood and asphalt shingle application on both new construction and reroofing jobs: How professional roofers make smooth tie-ins on any job, the right way to cover valleys and ridges, how to handle and prevent leaks, how to set up and run your own roofing business and sell your services as a professional roofer. Over 250 illustrations and hundreds of trade tips. **192 pages, 8½ x 11, $9.25**

Masonry & Concrete Construction
Every aspect of masonry construction is covered, from laying out the building with a transit to constructing chimneys and fireplaces. Explains footing construction, building foundations, laying out a block wall, reinforcing masonry, pouring slabs and sidewalks, coloring concrete, selecting and maintaining forms, using the Jahn Forming System and steel ply forms, and much more. **224 pages, 8½ x 11, $13.50**

Carpentry
Illustrates all the essentials of residential work: form building, simplified timber engineering, corners, joists and flooring, rough framing, sheathing, cornices, columns, lattice, building paper, siding, doors and windows, roofing, joints and more. One chapter demonstrates how the steel square is used in modern carpentry. **219 pages, 8½ x 11, $6.95**